BISCUIT JOINER HANDBOOK

• • • • •

Hugh Foster

 Sterling Publishing Co., Inc.   New York
Cassell PLC   London, England

Dedicated, with affection,

to my parents, who aren't even surprised when a project like this is completed. Now, that's faith.

to my wife and daughters, who complained less than they might have about the amount of time and energy this project consumed.

Edited by Michael Cea

Library of Congress Cataloging-in-Publication Data

Foster, Hugh.
    Biscuit joiner handbook / Hugh Foster.
      p. cm.
    Includes index.
    ISBN 0-8069-6800-1
    1. Woodworking tools. 2. Joinery. I. Title.
TT186.F67 1989
684′.083—dc19
                                               88-36591
                                                CIP

3   5   7   9   10   8   6   4   2

Copyright © 1989 by Hugh Foster
Published by Sterling Publishing Co., Inc.
387 Park Avenue South, New York, N.Y. 10016
Distributed in Canada by Oak Tree Press Ltd.
℅ Canadian Manda Group, P.O. Box 920, Station U
Toronto, Ontario, Canada M8Z 5P9
Distributed in Great Britain and Europe by Cassell PLC
Artillery House, Artillery Row, London SW1P 1RT, England
Distributed in Australia by Capricorn Ltd.
P.O. Box 665, Lane Cove, NSW 2066
Manufactured in the United States of America

Contents

Introduction

When the joiner first arrived on the United States market approximately a decade ago, it was easy to understand why many considered it just another expensive and extravagant gadget. Since then it has been refined, and its price, relative to the cost of labor, has become more reasonable. It is now available to and can be afforded by the home craftsman, with the result that the tool has become almost as essential to the serious woodworker as a cordless electric drill.

A plate joiner is not really used in the same way as other tools in the workshop. A well-known woodworker whom I know has had a Lamello Top joiner for years. He says that he'll often go most of the year without using it, but will then use a couple of boxes of biscuits—thin, elliptical wooden wafers that are

Illus. 1. Mitring a joint with a Lamello Top joiner.

used to join the slots cut by the joiner—within a week. This seems to be the case in my shop as well. But when you do use the joiner, you'll find that it is a very productive tool, primarily because it saves time.

Biscuits are approximately twice the price of commercial dowel pegs and markedly more expensive than shop-prepared dowel pegs. However, in today's workshop, material costs are just a small fraction of the project's cost. Woodworkers have to work quickly, neatly, and efficiently, all of which is accomplished with the joiner. It saves time, which translates to a savings in money. For example, I can cut dovetails by hand at a rate of about a foot and a half an hour (both pieces), and my cuts fit fairly well, so I waste little time during assembly. However, I can make a perpendicular biscuited joint that fits perfectly well in under a minute per foot and a half. And this biscuited joint is also an extremely strong joint. (See Illus. 3 and Chapter III.)

Illus. 2. A comparison of the three sizes of biscuits used.

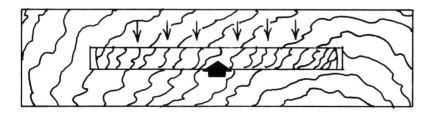

Illus. 3. This drawing compares the load-bearing capability of a dowel joint (top) to that of a biscuit joint (bottom).

If you're a woodworking student or teacher, you probably know that one of the perennial problems in a school shop is the amount of time wasted while students await their turns to accomplish each woodworking task. One way to expedite this process is to put four to six inexpensive plate joiners in the woodworking classroom and to teach students to use them rather than rely on older, more time-consuming methods. The joiner may well double the project output of the class and make the students more interested in woodworking—not to mention more employable as efficient woodworkers.

The joiner also has other advantages. Before I bought one, there were certain kinds of joints that I avoided making. For example, I very seldom used mitre joints, and never in a carcass application. Compound mitres were even harder to hold together, no matter how attractive they might have been. With the joiner it is now easy to make these joints.

Virtually all assemblies go together more easily when fastened with the joiner rather than with more conventional methods. The joiner will also make it more possible to assemble whatever design you have in mind, and will probably make the project more affordable for your customer. Illus. 4 shows a compact disc organizer that was assembled with no fasteners

Illus. 4. This biscuit-joined compact disc organizer took less than 15 minutes to cut out and assemble; if other joinery was used, this project would have been much more difficult to make. Note that it is hard to detect how the eight pieces are connected. Also note that the pieces should have been sanded better before assembly.

other than joining biscuits. Just a few years ago, this piece would have been far more difficult to make, much less produce in quantity.

Of equal importance with speed and productivity is the issue of safety. After hundreds of hours of experimenting with plate joiners, I am convinced that they are among the safest tools in the shop, safer than any of the other portable tools, including the sander. However, like routers, belt sanders, and most other electrically powered tools, plate joiners are loud. With the exception of the Porter-Cable model (see Chapter V), all of the joiners run just above 100 dB. (To put this figure in relative terms, remember that 100 dB [decibels] is the noise level for a very noisy factory, 65 dB that of normal conversation in a busy office, and 35 dB that of a quiet room.) So operators should wear some type of hearing protection.

Plate joining generates a lot of dust in shops. Some of the manufacturers discussed in the second section offer a fairly expensive dust collection attachment. If you are a left-handed operator, this option is really more of a necessity.

Illus. 5. A dust collection kit should be used on your joiner, which will otherwise deposit dust all over your work area.

There are a lot of advantages to a chip-extraction or dust collection system if you can justify its price. The handiest way to use it is to attach it to the power cord with either tie-downs or duct tape, and then put the extension cord that you're using to run the joiner near the intake on the vacuum.

In the first part of this book, the basics of joiner use are explored. This includes their history, how they work, how to use them, and a discussion of biscuits, which are, fortunately, completely interchangeable from brand to brand. In the second section, I examine all the commercially available joiners and their accessories, and discuss how to comparison-shop for a joiner. The final section deals with cutting techniques and covers the following: safety and maintenance procedures, procedures for making various kinds of joints—including butt and mitre joints—ways to adapt your machine to various nonstandard applications, and projects on which to try your skills.

One final note before you move on to the first section: For the sake of clarity, in the following pages the terms plate joiner and biscuit joiner should be understood to refer simply to a joiner. The thin, elliptical wooden wafers used for joining the slots cut by the joiner will be called biscuits, though when the manuals of the manufacturers are quoted they may refer to them as plates, splines, or wafers.

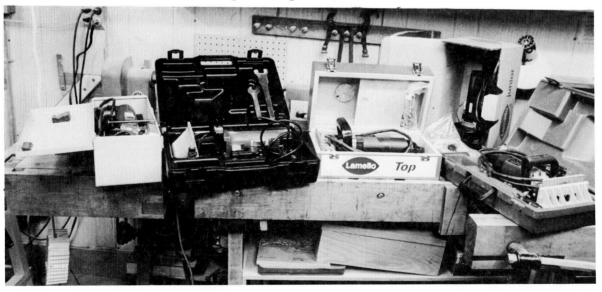

Illus. 6. Five joiners that are available on the market. These joiners and other joiners are discussed in depth in the second section.

Basics

I
Historical Overview

The history of biscuit joinery can be traced back to 1956, when Herman Steiner, a Swiss cabinetmaker, started manufacturing wood-joining plates under the brand name Lamello. (The name Lamello was derived from the German word *Lamelle*, which translates to "thin plate.") By 1969, the privately owned company had been transformed into Steiner Lamello Ltd., and began to manufacture a portable groove-milling machine. Four years later, the company had grown enough to move into a new plant in Bubendorf, 15 miles from Basel, Switzerland. By 1987, the company had grown to 40 employees and was building a second production plant.

Never content with the status quo, Lamello has constantly sought to improve its products. In 1974, it introduced dust-extraction equipment and a groove-milling machine with a pneumatic motor, and a year later the Lamello glue dispenser.

In 1977, the Lamello Top, similar to the machine that is bought today, was introduced. A year later K-20 clamping plates were marketed, and the following year the Lamello centering awl and the Mini-Spot G2 for patching resin galls were introduced. The Lamello clamping system was finally complete enough with clamps, tension hooks, and corner profiles to be marketed in 1980.

In 1982, the Spanish firm Virutex began fabricating the O-81 joiner, a high-quality fixed-angle joiner which has consistently sold for about half the price of the Lamello Top. As may be expected this tool has proved enormously popular in the United States and elsewhere.

Lamello's response to the Virutex O-81 came the following year with the Lamello Junior. Also marketed that same year were the Lamello Nova, used for trimming protruding edges in wood or plastic, and the Lamex assembly kit.

In the latter half of the 1980s, the following innovations were made to joiners: the combination right angle/mitre plate replaced the right-angle-only plate on the Lamello Junior; a spindle lock was added to the Lamello Top to ease blade-changing; and Freud introduced the first truly low-cost joiner, the JS-100. By early 1987, Freud's joiner had the old-line manufacturers scrambling to be competitive enough to keep their market share. By the middle of that year, even Freud was scrambling, first because its plants couldn't produce machines fast enough to keep up with the demand for them, and second because Porter-Cable introduced a radically different approach to joining that was also very cost-effective: a joiner that was driven by a belt rather than by helical gears. This type of joiner was shaped differently and made less noise than the joiners drive by helical gear, to name just a few differences. That same year, the Lamello 2000 stationary machine was introduced, as were C-20 joining plates for joining Avonite, Corian, Fountainhead, and other solid surfacing materials used in the kitchen industry. At that time, Kaiser entered the high-priced-joiner market.

It should be noted that though the history of joinery is short, the joiner has constantly undergone renovation to the point where it fills an important niche in the woodworker's workshop. Even as this book is being read, joiner manufacturers are working on ways to improve their products.

II

How Joiners Work

A joiner is basically a 4-inch die-grinder-like device with a special spring-loaded faceplate that sets the depth for plunge-cutting. (See Illus. 7 and 8.) Short spurs grab the workpiece while the blade plunges through the faceplate to make just the right cut for a biscuit, a flat football-shaped spline that is also

1. Central locking for cutters

2. Depth adjustment

3. Swivelling stop

4. Front with recessed grip

5. Clamping lever

6. Guide with surface finish

7. Antislip front

8. Inner chip guide with deflector

9. Handle

10. Switch

11. Spring

12. Base

13. Faceplate

14. Depth stop

15. Spindle Lock

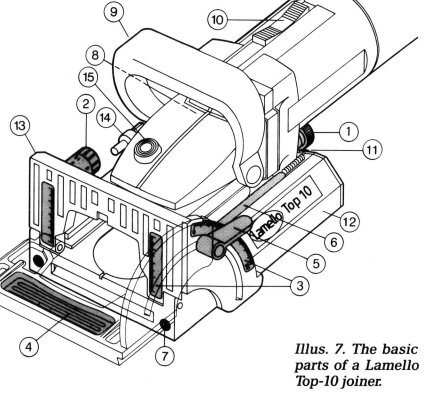

Illus. 7. The basic parts of a Lamello Top-10 joiner.

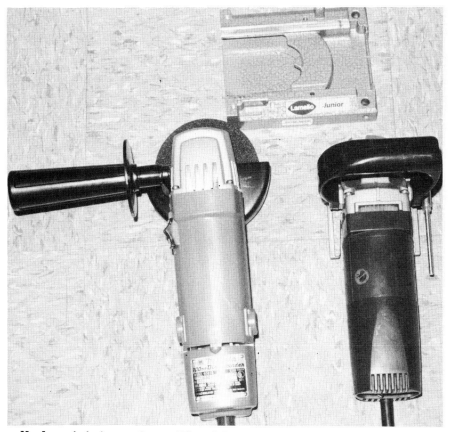

Illus. 8. Comparative view of a standard 4-inch grinder (lower right), a Makita grinder (lower left), and a Lamello Junior (top).

called a joining plate. The biscuits work like dowels and splines to help adjoining surfaces line up flush. (See the next chapter.)

The joiner provides a fast, easy, and accurate way to join wood in situations that might otherwise call for a mortise-and-tenon joint, a tongue-and-groove joint, or dowels. Though it is known for making quick, accurate butt joints, it can do much more. For example, with the joiner, edge-to-edge gluing joints and splined mitres can be made stronger and will be easier to align.

There are basically two kinds of joiners. The most commonly available kind—the Lamello, Virutex, Freud, and Porter-Cable models—plunges straight into its slotting cut. Another kind like the Bosch and Elu models pivots into its biscuit-slotting cut. This is basically a disadvantage, though this kind of joiner is able to make grooving cuts and make certain kinds of cutoffs more easily than the plunging joiner.

The spring-loaded centering pins, which are a great aid to accurate biscuit joining, help to keep all the non-pivoting joiners except the Kaiser (which uses a non-skid rubber face-plate) from moving as you make the plunge cut. These pins can be easily removed when you do other slotting work. Simply remove the faceplate by loosening two screws on the top and removing two screws from the face; the springs and pins will fall out. You can replace the faceplate for slotting to a maximum depth of $^{13}/_{16}$ inch; with the pins removed, the joiner does this much more handily than the portable circular saw.

Adjusting the Depth of Cut

Before operating the joiner, check, and if necessary, adjust its depth of cut. All the joiners that plunge straight in except the Lamello Junior and the Kaiser models have the same type of depth-of-cut scale. Illus. 9 shows the depth-of-cut scale on the Virutex O-81.

Illus. 9. The depth-of-cut scale on the Virutex joiner. This scale is representative of those on the other straight-plunging joiners with the exception of the Lamello Junior and the Kaiser models.

The depth-of-cut scale is marked for number 0, number 10, and number 20 biscuits, and you can quickly adjust it by simply pulling the depth-gauge plunger towards the front of the machine and twisting it to whichever position you desire before returning it to the working position.

On the Lamello Junior joiner, snap-in inserts are used to set the depth of cut. (See Illus. 10.) The Kaiser joiner has a cam-operated depth-of-cut scale that is very easy to operate, and is not connected to the blade housing, features that are discussed further in Chapter XI.

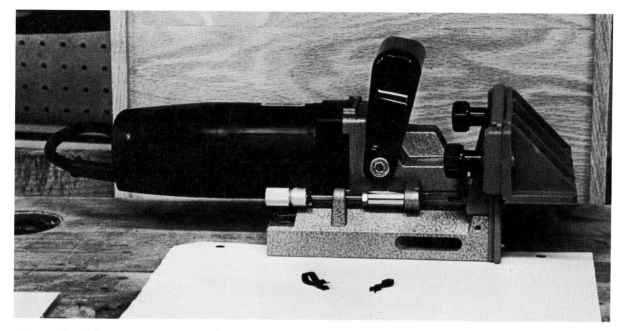

Illus. 10. Side view of the Lamello Junior. Note the snap-in inserts which are used to set the depth of cut for numbers 20, 10, and 0 biscuits.

The depth-of-cut scale on joiners that plunge straight in can be fine-adjusted with a pair of knurled jamb nuts. Make this adjustment every time after you remove the blade cover, and check it periodically.

On the angle-in joiners, adjust the depth of cut by tightening or loosening the depth-of-cut screw to specific places on a scale. A clockwise adjustment makes the depth of cut shallower; a counterclockwise rotation makes the cut deeper. Illus. 11 shows the adjuster on the Elu model 3380.

With either type of joiner, it is easier to set the depth of cut with a number 20 biscuit that you have sawn in half lengthwise rather than with a ruler. After you have set the depth of cut, you are ready to cut all three size biscuits. Adjust the depth of cut with the plunger, and occasionally check the fine-adjustment setting.

Illus. 11. The depth-of-cut scale on the Elu joiner, an angle-in joiner.

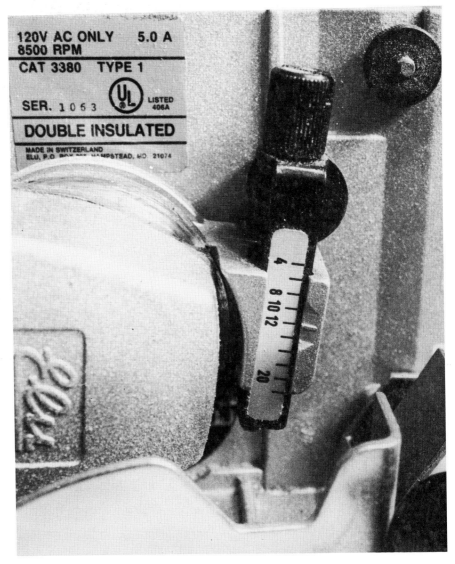

Making the Joint

To make a joint, mark out the pieces to be joined two inches from either end and about four inches apart between the outer two marks. This can be done with a scale or by eye; after you have used the machine for approximately an hour you will be able to mark out quite handily by eye. The marking out does not have to be elaborate.

Accurate setup is the key to accurate work. Cut out the

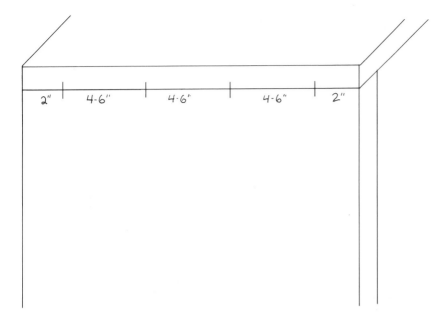

pieces so that they will fit well when they are assembled. Take extra care when you cut them; this meticulousness will be rewarded when you begin joining them.

After you have cut out the pieces, proceed to mark out the joint. Mark to the sides of the joint rather than to its center. Simple pencil scribes made with a template, ruler, or by eye are adequate.

Lay out the joints logically. Since you are marking the sides of the joints, use the same side all the time. Mark clearly which pieces go where. Don't rely on your memory.

To cut the joints, set up the spacing of the cuts with the work and the tool both flat on the bench; this position allows you to center the cut in ¾-inch stock (the most commonly used wood thickness) and permits bilayered joining in thicker stock. If you need more than two layers further adjustments can be made.

Though there are "square" guides on each joiner, they are not always enough to guarantee that the joint will be square, so a "stacked" setup gauge similar to that described in Chapter VIII is advisable. (See Illus. 13.) If the slots on the mating pieces aren't exactly parallel, the pieces will not join accurately or at all. In some cases, it is more than the middle that is out of alignment. For example, a ³⁄₃₂-inch miscut turns out to be a ³⁄₁₆-inch error when the pieces are joined, so the joint obviously won't work.

Illus. 13. It's a good idea to make a set of setup blocks like the one shown on the bottom of this photo. They only have to be 6–8 inches long, 2–4 inches wide, and the thickness of the stock that's most commonly worked.

Whichever way you cut the joints, they glue up alike. Glue carefully. Gluing inside the biscuit slot, as shown in Illus. 14, is usually sufficient except when you are using biscuit joinery to align the board for edge-gluing.

Illus. 14. This is all the glue that most biscuited joints require.

Illus. 15. A Lamello glue bottle was used to glue together this small carcass unit; the total joining time on it was less than five minutes. The Lamello glue bottle is discussed in Chapter XIV.

Illus. 15 shows a box with a divider. Let's discuss the procedures for gluing it. Lay one of the pieces that will carry the insert on its side. Glue only the slots; run a fair bead of glue down each side of the slot or use a special biscuit-joining glue bottle. (See pages 89–91.) Insert the biscuits in each slot and then glue the biscuits on the pieces to be attached. Attach them immediately. Apply the next batch of glue and biscuits, and finish the assembly. It takes longer to describe the process than it does to do it.

It is a good idea to test-fit the piece with dry biscuits before gluing. To accommodate the glue, cut the grooves a bit deeper than half the biscuit's width; this also makes them slightly longer than the biscuit, thus allowing nearly a quarter inch of lengthwise play so that you can adjust the pieces to be flush at the ends.

A joint fastened with biscuits is extremely hard to tear. After it has been glued and clamped for just ten minutes in warm weather with a fast-set glue, it might be impossible. Chiselling firmly planted biscuits is extremely difficult, even with a razor-sharp chisel, so plan carefully.

EDITOR'S NOTE: *Biscuit Pucker*—the author's response for *Fine Woodworking*'s "Ask the Experts" column.

I've seen the origins of this problem on one occasion: When using ⅜-inch stock the biscuit filled the kerf so completely that its expansion forced biscuit-shaped protrusions to appear on both faces. When you glue an assembly, if you scrape the glue bead off before it's dry, and belt-sand the work to make it flat and clean, you may never notice these slight protrusions. But, as the wood adjusts, the protrusions that you've sanded off shrink to leave puckered indentations. Keeping the biscuit ¼ inch from any surface is a satisfactory solution. In five years, I haven't had this problem, and most of my joinery has been in ¾-inch stock. You may get in trouble putting two rows of biscuits in ¾-inch stock; one is enough.

To keep glue from sticking to the biscuits, a coating of lemongrass oil or silicone should work. But, while preventing the expansion that may lead to biscuit pucker, it largely defeats the purpose: The swelling is what makes biscuits so efficient.

III
The Joining Biscuit

Biscuits come in three different sizes. A number 0 biscuit (⅝ × 1¾ inches) is 8 millimetres wide. A number 10 biscuit (¾ × 2⅛ inches) is 10 millimetres wide. A number 20 biscuit (1 × 2⅜ inches) is 12 millimetres wide.

As measured with a surface gauge, each biscuit is .148 inch thick, and though thinner than the 5/32-inch (.156-inch) saw blade that cuts the slots, it swells rapidly with moisture to .164 inch, enough to grip the slots tenaciously.

When the biscuit is inside the slot, the result is a joint that's very sound mechanically, which is essential. In fact, a biscuited joint after just 20 minutes of clamping is stronger

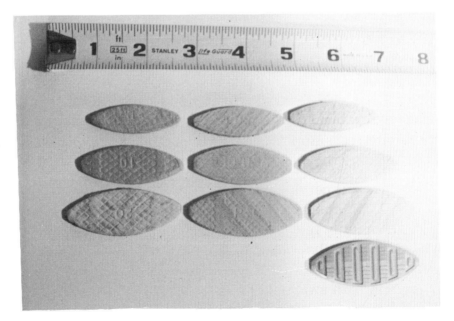

Illus. 16. A look at some of the commercially available biscuits. On the left are generic biscuits, in the middle are Porter-Cable biscuits, and on the right are Lamello biscuits. At the bottom right of the photo is a K-20 clamping biscuit. All these biscuits are very similar, so buy the cheapest ones available.

than a dowelled or splined joint and as strong as standard mortise-and-tenon joints, which are harder to make.

Though some companies insist that their brands of biscuits be used with their joiners, any brand of biscuit will work with any brand of joiner. Illus. 16 compares generic, Porter-Cable, and Lamello biscuits, and a Lamello clamping biscuit, which is shown for size comparison.

When first starting out, buy one package of each size of biscuit. Keep a spare package of the least-expensive brand of number 20 biscuits. You will find yourself using lots of biscuits. (See Illus. 17.) And if you do not live in the city, you will most likely not be able to buy the biscuits from your local hardware dealer. I telephoned my first order for biscuits to the company that advertised the best price; that company supplied biscuits within a week, which was good service. However, if you need biscuits desperately, a week can be too long.

Illus. 17. A stack of biscuits like the one shown here is required for even a small project. It's a good idea to have a few cartons of biscuits on hand.

When you get the biscuits home, store them in sealed containers (for example, Ziplock bags) because they are affected by humidity. Improper storage could be expensive, since the biscuits cost about 3 cents apiece. Some biscuits fit very tightly when they are first inserted and if they are stored improperly they can swell and become very difficult to insert.

After reading articles about drying carving stock in a

microwave oven, I considered the possibility that perhaps biscuits that have swelled can be reused after they have been microwaved. Perhaps it is a matter of figuring out how much moisture comes out of the biscuits in a certain period of time at a certain wattage, and determining how much the biscuit shrinks. Then, after making the calculations and microwaving the biscuits, the next step would probably be to compress them in a vise or strike them with a hammer to return them to a dry size so that they can be used. Perhaps it might be easier to go over them lightly with a disc sander.

After all these steps, would the biscuits work as well the second time? Since a half box of biscuits represents an expense small shop owners can't overlook, I decided that it was worth an hour of experimentation to find out. After using a biscuit to set the dial indicator on my moisture metre at zero, I wet a biscuit and watched it expand by .015 inch over a period of approximately ten minutes; then I microwaved it for 90 seconds on "high," moving it every 30 seconds. Each time I moved it, there was a biscuit-shaped "puddle" on the bottom of the oven, so moisture was obviously coming out of it.

After ninety seconds the biscuit was quite warm, so I remeasured it. All but .003 inch of the swelling had vanished, even without compression. Rewetting it brought the biscuit back up to .012 inch almost instantly.

The experiment was a tentative success. Illus. 18 suggests that biscuits so treated might be fragile (I was able to break this biscuit by hand, and I could not do that to one that hadn't been swollen and reshrunk) compared to other biscuits. Keeping biscuits dry in storage is the best plan of all.

Illus. 18. This broken biscuit snapped very easily after my dehydration experiment.

Biscuit Strength and Effectiveness

The first set of sample joints generally convinces most wood-workers of a need for a joiner in their shops. Here is a test that will help you determine the strength of biscuited joints: Make a small panel by taking two small pieces of hardwood, jointing them, and cutting joining slots two inches on center from either end, and another halfway between them. Lay a fair amount of glue down the center of each piece; be sure to get some into the slots you've sawn with the joiner according to all the manufacturer's instructions. Slide in three biscuits and clamp the piece with a single clamp. After allowing for the proper clamping time, you'll be surprised by the strength of the joint.

When I conducted this experiment in a shop that was heated to only about 60° F, I put the clamped piece in my office so that the glue, ordinary Titebond, could dry. After only ten minutes of drying time, I brought the piece back into the shop, scraped off the excess glue (which still had not hardened), and ran the piece through the surface planer on both sides. As I "stressed" the piece in my hands, I could see a little glue working in and out of a snipe in one end of the joint, a product of our hasty stock preparation. Of course, the glue hadn't set in 10 minutes at that temperature and the joining plates were all that were holding the panel together.

Wondering what it would take to break the joint, I began to "torsion" the panel in my hands; After I put stress on the board for five minutes at full strength, the joint broke—but none of the three biscuits broke. The glue inside the joint was still wet.

Illus. 19–23 attest to the strength of joints made with joiners. Illus. 19 shows two small walnut pieces that were joined with one biscuit, clamped overnight, and then sawn open to reveal the mechanical nature of the joint.

Illus. 20 shows two pieces of maple, each $1 \times 3 \times 12$ inches, edge-joined with three biscuits. These pieces were scrap from my very cold winter shop. I was impatient to glue the machined pieces when I brought them into the house from the shop and set them on the radiator for nearly an hour, a foolish move. The pieces began to check badly, but I glued them anyway and let the glue dry overnight. Despite the serious

Illus. 19. These two small pieces of walnut were biscuited together and then sawn open to reveal the mechanical nature of the biscuit joint.

Illus. 20 (above left). These boards began to check badly after they dried out on a radiator while being warmed for gluing. Note, however, that the biscuits did not fail, a tribute to their strength. Illus. 21 (above right). To break the biscuit that joined these ¾ × 2½ × 30-inch pieces, I had to put one piece in my bench vise and apply all of my body weight to the other member.

checking, some of which appeared to run two-thirds of the way through the boards, I had to beat the pieces with a hammer to break them, and even then it wasn't the joint that broke!

Illus. 21 shows a butt joint that was sloppily made with a single biscuit; you can see where I cut through the side slightly. The joint was biscuited and glued in place without clamp pressure, and left overnight. The joint would not break under just hand pressure, but only after I put one member in my bench vise and applied full body pressure (about 190 pounds) to the other.

Illus. 22 and 23 show close-ups of broken joints made in particleboard. Illus. 22 reveals that the piece of sheet stock

Illus. 22 (above). A broken joint in particleboard. Note that the biscuits have remained intact while the sheet stock has been destroyed. Illus. 23 (below). Another broken biscuit joint in particleboard. Note that once again it was the particleboard that broke under great pressure, not the biscuit.

broke before the joint did. Illus. 23 shows how much of the sheet stock clung to each biscuit in another forcibly broken joint. A lot of modern furniture is made of veneered sheet stock like this.

If you use biscuits, you will be virtually guaranteed an extremely solid joining work. Noted furniture-maker Graham Blackburn joins chairs with biscuits. Even load-bearing shelves such as those used for phonograph records can be biscuit-joined in place.

How to Use Biscuits

All joining biscuits are made of solid beech. According to Lamello, its joining biscuits are made as follows: felled tree trunks are cut to length and sawn into boards, which are then cut to squared timber and dried. Then the material is sawn into laths, which are processed to plates on a stamping press before they are sorted, counted, and packed.

Lamello claims that its biscuits are evenly pressed and feature fine-ground edges with ends blunted to allow easy escape of superfluous glue. The biscuits manufactured by other companies do not look much different to me except for the amount of the ends that has been cut off. Porter-Cable claims that its American-made biscuits have more of a bevel on all their edges to allow for easier insertion, but, once again, I have a hard time detecting the difference.

Biscuits are used for joining surfaces, corners, and frames, and can be butt-joined, staggered, or mitred. They can be used on chipboard, solid wood, plywood, or other man-made sheet materials.

Biscuits work like splines and dowels as they help to line up adjoining surfaces. However, they are preferred over dowels because they provide a greater wood-to-wood surface gluing area, though you should use as many of them as possible since they are the only source of structural integrity.

When using the biscuits, simply do the following: Align your cuts. Since the biscuit slots are cut slightly larger than the biscuits are, you don't have to line up your cuts perfectly lengthwise. (See Illus. 24.) However, the joiner must be set up square or it may misalign things widthwise.

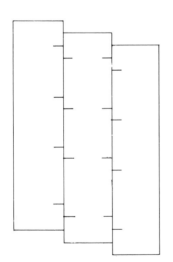

Illus. 24. You have about ¼ inch of leeway when you are aligning a pair of joints; this should be enough to allow for perfect alignment.

Illus. 25 (left). Use a knife-like implement to remove the chips from the slots before beginning critical assemblies.

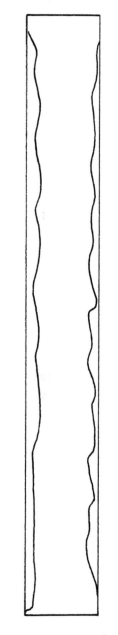

When the slots are properly aligned and cut in (Illus. 25), glue the slots with PVA (polyvinyl acetate) glue like Elmer's or Titebond. Glue carefully. Dripping the glue down both sides of the slots (Illus. 26) is the best way of gluing except when you are using biscuit-joining to align boards for edge-gluing or when you are joining mitres. Just filling the slot can be too messy, and gluing the biscuit would dictate assembly times that are impossibly short.

The biscuits will get wet soon after insertion and expand, thus producing the required lateral pressure inside the groove. The continuous setting process of the glue leads to growing mechanical strength, which means that only a very short time is needed for clamping.

While biscuit-joining itself is very fast, take your time when cutting the members to be joined. Cut them very accurately and plan the assembly of your project carefully. A biscuit-joined project can be assembled much more neatly than other types of projects. You will find that you use far less glue, so the pieces can be all but "finish-sanded" before being assembled.

Illus. 26. How the glue should look in the slot.

Accessory Biscuits

The plastic, toothed size-20 Lamello clamping biscuits are expensive. A box of these biscuits, however, should last a long time. You may not use them often, but when you do, you'll realize how valuable they are. These biscuits, shown in Illus. 27, should be popular with all plate-joiner users. They are timesavers on projects that are too awkward in shape to clamp or on projects where the neat gluing that one gets with the Lamello glue bottle isn't even good enough. I installed one of these halfway, changed my mind, and then tried unsuccessfully to remove it with a pair of pliers. This attempt, shown in Illus. 28, destroyed the biscuit before it came out, and the attempted removal did the piece of particle board no good whatsoever. Need I even mention that I decided to leave future K-20 biscuits in place?

Illus. 27. Lamello K-20 plastic clamping biscuits are used with wooden biscuits for hard-to-clamp joints. Available only in size 20, a carton of these biscuits may last quite a while in your shop and help you to complete awkward joining projects.

Illus. 28. A K-20 clamping biscuit after it has been broken out of its slot. Note that the pliers and the 1/8-inch chisel required to remove the biscuit not only broke the biscuit into many pieces, but they also damaged the slot.

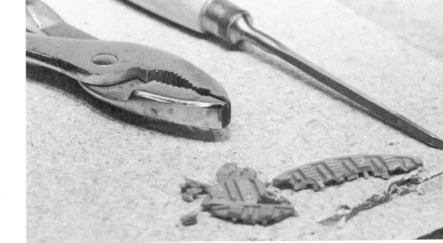

Illus. 29 shows a similar product, the newly issued C-20 joining plate made exclusively for joining Avonite, Corian, Fountainhead, and other solid surfacing materials that are becoming very popular for counter tops. These joining plates are made of clear plastic.

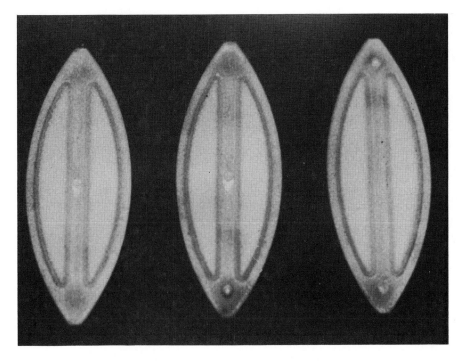

Illus. 29. Lamello C-20 joining plates are used to join solid surfacing materials that are being used increasingly for counter tops.

Lamello Simplex knockdown fittings, as shown in Illus. 30, are made of aluminum and are expensive. In the right application, they could be very useful. Sooner or later, we will be buying these fittings for frames, aprons, and for other supports for heavy loads. One can even use them in tandem on a bed frame.

As with the other Lamello accessories mentioned here, you don't have to plunge-cut with a Lamello brand-plate joiner to be able to use the fittings. Installation is easy: Simply mill mating slots, apply epoxy cement in both slots, and insert a connector in each slot. Lamello even has an insertion tool, shown in Illus. 31, for this job if you want to ensure dead-centered positioning every time. This tool is certainly more of a luxury than a necessity—I had great success installing the connectors free hand.

Illus. 30. Lamello Simplex knockdown fittings are made of aluminum. Sold in pairs, they are very helpful on items that must be quickly assembled or taken apart.

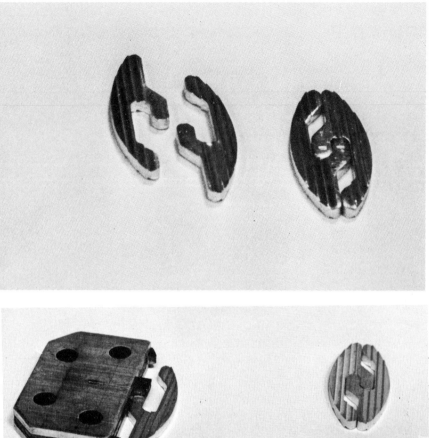

Illus. 31. A Simplex insertion tool and a pair of inserts.

Commercially Available Joiners and Accessories

IV
Buying Guidelines

The following chapters in this section describe joiners and commercial accessories available in the United States as of early 1988. Manufacturers are continually making changes and improvements, so get up-to-date descriptions and price information from them. Appropriate addresses and a summary of each unit's features are provided in Appendix B on pages 188 and 189. Also refer to Appendix A for an update on the latest joiner improvements and accessories.

Don't buy a joiner sight unseen from a catalogue photo or a description in an advertisement. Don't even accept verbatim my descriptions of the models. Operate the joiner or at least examine it firsthand before deciding which model to buy. You'll have to live with this purchase for a long, long time.

The joiners available today share many of the same features. Many manufacturers, particularly those whose equipment costs more, do not like to admit this. Illus. 32 shows an interior view of the Lamello Top joiner, and Virutex and Freud joiners. The insides of these machines appear very similar to the untrained eye, but the tolerances of the machined parts get progressively finer as the units become more expensive. I won't attempt to tell you whether the finer machining can justify the joiner's price: Only you and your credit card can make that judgment.

Until 1987, plate joining was expensive. Available at that time were the two Lamello machines and the Elu model. While these machines are still among the very best, they are the most expensive today—even with the deep price cuts brought on by the competition, Freud and Virutex. The competition has been able to sell their machines at lower prices than these manufacturers partly because they have been able to take advantage of the research and development put in by these companies. Illus. 33 shows all the joiners available in early 1987. The additions to the market since then have enhanced the selection you have to choose from.

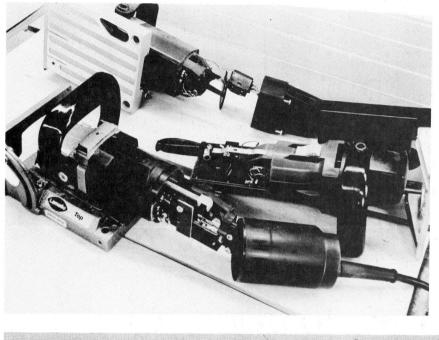

Illus. 32. Shown from front to rear are interior views of the Lamello Top, Freud, and Virutex joiners.

Illus. 33. Shown from left to right are the Freud, Virutex, Elu, Lamello Top, and Lamello Junior joiners. These joiners and other models will be discussed in the following chapters.

What happens if (when) your joiner breaks down? After you've stopped cursing, read your warranty. Good help is available with whichever machine you choose. The newest units available—the Porter-Cable, Freud, and Kaiser models—offer a full year's "limited" warranty, as do Elu, Bosch, and Virutex. Freud is the only firm with a toll-free help line. The Lamello joiners, considered by some to be the very best, are warranted for six months. In any case, the best warranty is to ensure that you will not have to depend on one.

If after reading the following chapters you determine which joiner you want, there are a few shopping guidelines you should abide by when buying your unit. Remember to operate or examine the unit firsthand. Also, I believe it's unethical to shop locally to get the feel of a tool, only to mail-order it from a discount store that can beat the local merchant's price. If you use the local dealer, you should buy your joiner from him or there may not be someone to supply that information next time you go shopping. After all, your purchases provide those dealers' incomes. Now, I'm not saying that you have to buy locally—indeed, you may not be able to afford to—but it's not fair to expect the local merchant to provide the services when a distant dealer gets the profit.

When you get that joiner (or any other tool, for that matter) home, write the model number, serial number, and date of purchase on the cover of the manual. Then punch holes in the manual, and keep it with all your other tool manuals in a ring binder; my ring binder also includes sales receipts, repair records, and other related material. Keeping your materials organized thus ensures that you'll have them when needed. Sooner or later, even the best tools need repair. They weren't made to last forever. Keeping the necessary materials organized is but the first step in prolonging the tools' lives.

One quick word about Asian manufacturers. So far, none has introduced a joiner, or has even been willing to indicate that one is in the works. When the Asian tools reach the market, if they do, it will be greatly transformed.

Now, on to the individual reviews.

V
Porter-Cable 555
Plate Joiner

The Porter-Cable 555 is the first, and so far only, American-made entry in the joiner market. This joiner has several features that make it unique. Its 5-amp, 8,000-RPM motor drives its blade by belt rather than by helical gear, so it's by far the quietest of the units. (See Illus. 35.) With my sound-level meter, I gauged its noise level as 95 dB with no load, and 93 dB under load; under load it is much quieter than the next least-noisy joiner.

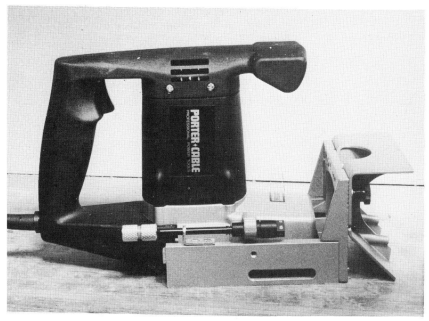

Illus. 34. This side view of the Porter-Cable 555 shows the workings of the fixed-angle faceplate and the depth-of-cut gauge.

This belt drive is not only quiet, it is also strong. The joiner will supposedly cut 20,000 slots in oak without produc-

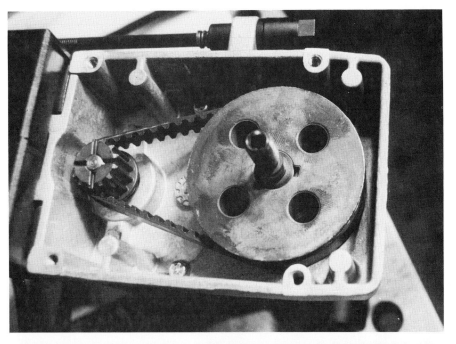

Illus. 35. The Porter-Cable 555's drive belt is unique among the joiners.

Illus. 36. The inside of the Porter-Cable 555 joiner. Note the "super torque" belt.

ing any noticeable wear on its toothed, "super torque" belt. While it might be a good idea to have an extra belt on hand, my own extensive use of the tool has produced no noticeable belt wear, so a spare is not really needed for the first couple of years of use.

Because of its radical design, the Porter-Cable 555 is not only one of the most comfortable of the joiners to operate, but at just under five pounds it is also the lightest. One could easily operate the tool all day. I have on many occasions.

Because it is so quiet and light, the Porter-Cable 555 is the safest joiner to use. These features reduce stress on the operator and those working around him.

The joiner's 7-foot cord is adequate, and its American-made metal case, shown in Illus. 37, may well be the sturdiest of the joiner cases.

Illus. 37. The Porter-Cable 555 in its case.

The manual is a model of Porter-Cable's usual excellence in communication. In addition to the safety precautions that are common to all tool manuals, there are items specific to joiners that are worth reiterating here since so many of the other manuals omit them. They are as follow: The guard base must be kept in working order. Check its operation before each use, and do not use the tool if the guard does not close briskly over the blade. If the joiner is dropped, the guard may distort, thus restricting operation and making the tool dangerous. Keep the slide mechanism free of wood chips. Occasionally lubricate the ways with light machine oil, as this prevents excessive sawdust buildup. Keep the blades clean and sharp, for sharp blades minimize stalling and kickback. Guard against kickback; release the switch immediately if the blade binds or the saw stalls.

Porter-Cable is the only manufacturer with a manual that recommends practice cuts. I find this honesty refreshing!

The Porter-Cable 555 is different from the other joiners in that it indexes for mitre cuts against the outside of the work rather than the inside; this is what causes the Porter-Cable joiner to look radically different from other joiners. Illus. 38 shows the unusual faceplate.

Illus. 38. The faceplate on the Porter-Cable 555 is different from the faceplates on the other joiners.

Mitred work should be laid out against the inside of the joint; the layout line corresponds to the long line down the center of the tool's bottom. (See Illus. 39 and 40.) While seeing the work in this position may tempt you to operate the joiner upside down while squeezing the trigger with your little finger

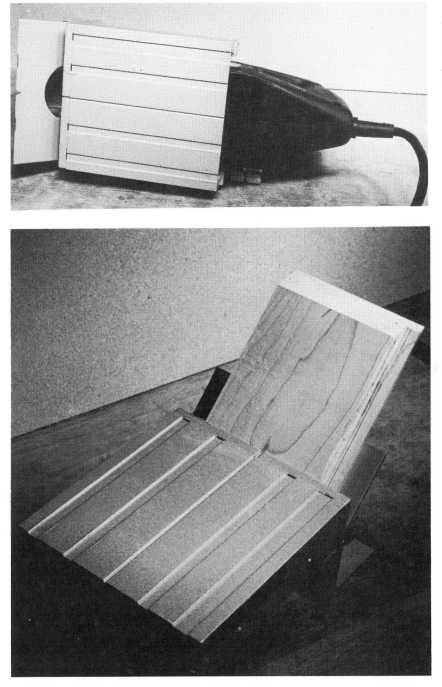

Illus. 39. The underside of the Porter-Cable 555. Note the long layout line right up the center.

Illus. 40. The marked-out piece resting properly in the mitre-cutting gauge.

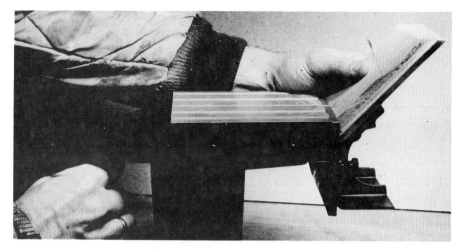

Illus. 41. For the sake of accuracy and safety, avoid cutting mitres the way that is shown here: with the Porter-Cable 555 resting on its top while you operate the trigger with your little finger.

(Illus. 41), avoid this technique for the sake of both accuracy and safety. While the operation may be safe enough with a joiner, it could lead to unsafe habits with other tools, and safety should always be a consideration in the shop.

Illus. 42 shows the proper way to cut mitres. Mount the piece to be mitred in a sturdy vise. Thus, as Illus. 43 shows,

Illus. 42. The proper way of cutting mitres. The piece should be held in a sturdy vise.

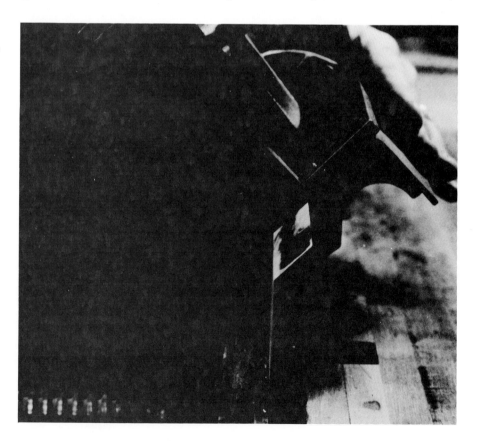

when pieces of unequal thickness are mitred they meet correctly at the outside edge of the joint rather than at the inside edge.

Because all other joiners join mitres by gauging from the inside of the joint, this sometimes results in either gaps in mitre joints or in the operator cutting the biscuit slot so deep that after sanding there are joining "blemishes" on the surface. Of course, you may run into the same kind of problems while using the Porter-Cable joiner (Illus. 44) if you do not set it up properly.

The inverted 45-degree fixed-angle faceplate shared a spot on the prototype with a flap-front faceplate similar to the one found on the Lamello Top. (See Illus. 45.) Illus. 46 seems to indicate that it might be possible to mount a flap-front faceplate to the production model of the Porter-Cable 555. If it were, the results would be an intriguing machine that would, of course, cost much more. If enough users requested it, Porter-Cable might be tempted to make a machine like this. After all, it was only after the pre-production models were tested and suggestions made that the layout lines on the Porter-Cable 555 joiner were made longer and narrower.

Unlike other manufacturers, Porter-Cable doesn't include a small oil bottle with its joiner; this oil bottle helps to remind joiner users to lubricate the slides. Even though the sliding platform on the joiner fits extremely well, it will be necessary for you to occasionally lubricate it lightly. Lacking

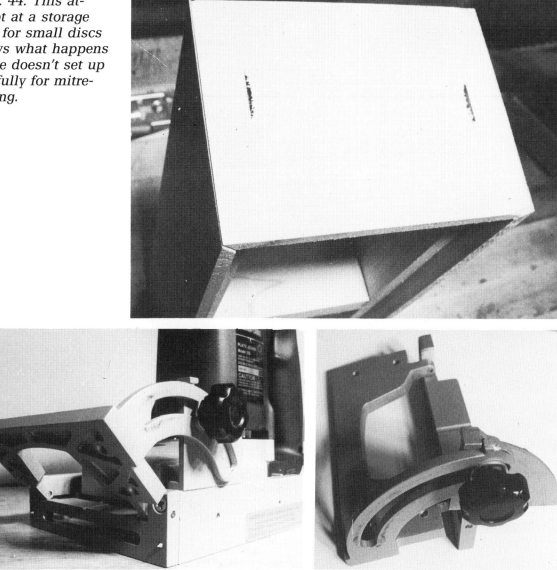

Illus. 44. This attempt at a storage case for small discs shows what happens if one doesn't set up carefully for mitre-joining.

Illus. 45 (above left). The flap-front faceplate mounted on the prototype of the Porter-Cable 555. Illus. 46 (above right). A closeup look at the flap-front faceplate. The Porter-Cable 555A features a flap-front faceplate—an accessory offered by Tools, etc. of Fullerton, CA—that is much like the one shown here.

the manufacturer's sample of lubricant, I have been using TRI-FLO spray lubricant with great effectiveness.

Though only mitring cuts have been addressed in this chapter, the Porter-Cable 555 is capable of making any joint that requires biscuits. Illus. 47 shows this machine being used to cut a T joint. This joiner offers an opportunity to quickly and

Illus. 47. The Porter-Cable 555 being used to cut a T-joint.

efficiently make all types of joints, increase your efficiency as a craftsman, and perhaps achieve a new, higher level of quality in your work.

A note should be made here about Porter-Cable biscuits, which are also American-made. They are tapered for easier insertion and have notched ends. When you combine these features with the fact that the biscuits are considerably cheaper than the next least-expensive brand of biscuits, you can easily understand why they are so popular.

VI
Delta 32-100
Plate Joiner

The first stationary biscuit joiner comes from Delta International's Taiwanese operation. This joiner should help to enhance the reputation of Taiwanese tools, which has suffered from many tools appearing to be poorly made. The Delta 32-100 joiner will go a long way to changing that perception (Illus. 48): The castings are nicely made; where pieces are supposed to be fitted together by the user, they fit snugly; where parts are supposed to move against one another, they do so smoothly; and the tool appears to have been made convenient and pleasurable to use rather than merely inexpensive. I have seen it advertised for much less than its list; especially at the lower end of this price range, it represents a tolerable bargain.

Unlike all the other biscuit joiners discussed here, this one is designed to be mounted to a bench. I made a "portable" bench for mine. While in my shop this portable stand served only until a permanent stand was made; it doesn't need to be temporary and seems an excellent idea for shops that want the convenience of the stationary joiner but are chronically short of space. The only possible problem with the portable stand might be kinking the drive wire.

To make a base so that the tool can be clamped to your workbench when you want to use it and put elsewhere when you don't, use a 1 × 8¾-inch by the appropriate length; on my workbench, 20 inches happens to be perfect. Mark out the board so that ¾-inch countersinks ¼ inch deep can be drilled 6 inches (all these dimensions are "on center") apart in the board's width (starting 1¼ inches from the front) and 9¾ inches apart lengthwise. (See Illus. 49.) After the joiner is mounted, the board must be clamped down before the tool is used. On my considerably more than three-foot-high bench, this makes the cutting height of the tool 44½ inches. On my

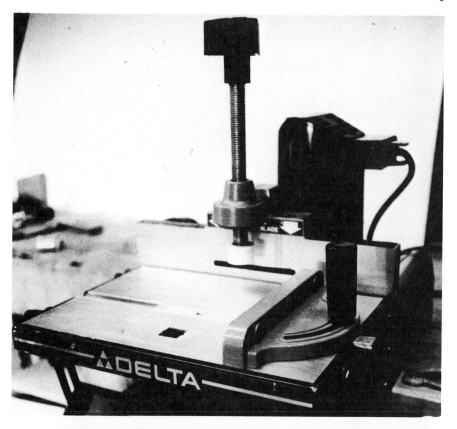

Illus. 48. View of the Delta 32-100 plate joiner machine and worktable with clamp in place for horizontal work—and with bevel-cutting table hanging below.

Illus. 49. The shop-made base clamped in place on the bench before the 32-100 is attached to it.

permanent stand, the cutting height will be *at least* 8 inches lower; after all, convenience is what this tool is about.

The Delta 32-100 is loaded with convenient features. There are quick-release threads on the height-adjust and the hold-down screw. A fine-adjust knob will make the depth of cut exactly what you want. (See Illus. 50.) These convenient features are added to the ultimate convenience: accuracy. The tool travels squarely through these positions. Delta advises the operator to tighten the right (scale-side) knob first to keep the unit's travel square.

What looks like a mitre gauge is actually an adjustable stock stop which can be screwed into place from the underside of the table and positioned anywhere on the table, with either right- or left-hand reference. If you adjust this regularly, then you won't want to leave the angle guide just hanging under the main worktable (as the owner's manual recommends for much of the time).

Illus. 50. View of the fine-adjust knob on the height control.

Illus. 51. Set up a bevel cut by firming the edge to be cut squarely against the tool's face; another surface will be clamped to the bevel-cutting table.

A steel tilting table is provided for joining work with bevelled edges. (See Illus. 51.) This is far handier than all but the flap-fence of the Lamello Top-10 (See Chapter XIII). And since the work can still be clamped to the table, this may be more accurate. Note that the levers that lock the tilting table in place are spring-loaded, so that they can easily be positioned out of the way.

The tool measures 89 dB from three feet, and 92 dB when plunged into air, but only 90 dB when plunged into a piece of hardwood that is clamped to the table. A ribbed belt (like the Super-Torque belt in the Porter-Cable 555) transfers the power from the motor to the cutter. (See Illus. 52.) A spring returns from the plunge-cut without jarring the machine even slightly. A 12-inch-wide machined face is nicely marked for size 20, 10, and 0 biscuits.

The joiner is 19 inches deep, 14¾ inches wide (including the items that overhang the 12 × 8-inch table). Overall, the machine is 15 inches tall.

An added safety feature is that the tool can be locked "off" simply by installing a padlock through the switch block.

The clamp/hold-down is so easy to use that the operator may actually use it! The adjustable table and the machined face have both been designed to accept its tight-fitting square base for clamping both horizontal and vertical work. Clamping work firmly to the table is one of the secrets that leads to this unit's very accurate work. On some bevel cutting, you may prefer to use another kind of clamp to hold the work down, such as a C-clamp. Additionally, when cutting bevels, it is good practice to clamp a backup stop block to the angle fence to take the place of Delta's adjustable stop stock on the regular table.

The cable to the foot pedal may be the weakest link in

Illus. 52. The underside of the main casting showing the ribbed belt that drives the unit.

Illus. 53. View of the 32-100 mounted on the workbench with the foot switch that operates the cutter's plunge action.

the system. (See Illus. 53.) The instructions are adequate for assembly, and it works perfectly well, but it is prone to kinking and might be hard to replace afterwards.

The scale on the side of the unit measures from the bottom of the opening rather than the bottom or middle of the cutter. So, to place a biscuit properly on the middle line in ¾-inch stock, one must interpolate ⁷⁄₃₂ of an inch on the scale. Alternatively, the fine-adjust knob makes it awfully easy to line up marked work.

Layout is even easier with this machine than with a standard biscuit joiner, especially if you have repeated slots to cut. In that case, mark out the first piece, and clamp the board and the guide; after the first cut is made, it is just a question of placing a new board against the guide to facilitate repeat cutting.

Operations are standard, but the Delta's face shows the exact width of cut as well as a middle line. If the boards to be joined are not the same thickness, they must be positioned on the joiner "good" side down.

When joining bevels, it is best to have both pieces bevelled at the same angle. If, for some reason, they aren't, the biscuit slots must be cut at the same angle, since the biscuits won't "flex" to fit.

Butt joining is a good exercise in using both horizontal and vertical clamping. (See Illus. 54 and 55.)

Even more than with the hand-held joiners, the operator has to pay attention to the angle fences; when I first mitred on the Delta 32-100, I mitred *through* the stock, ruining it, and harming both the blade and the mitre table.

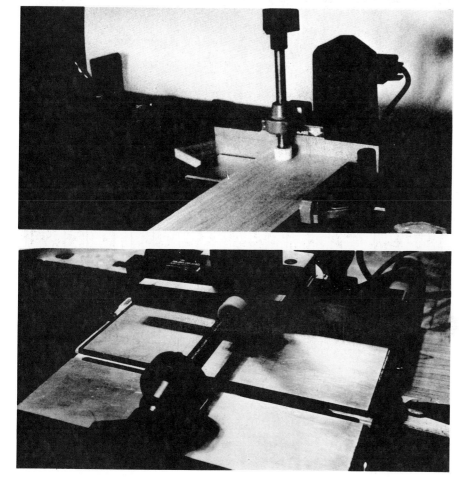

Illus. 54. Setup for horizontal work.

Illus. 55. A protective clamping block is a good idea when setting up either vertical—as in this case—or horizontal work for cutting.

Illus. 56. A finished mitred edge with groove and biscuit sits on top of the mitred edge to be cut while held in the two-way clamp.

Illus. 57. A view of the framing-type joint that the 32-100 cannot make.

Cutting picture-frame mitres is simplified with the layout aids of the 32-100 joiner. (See Illus. 56.)

Edge-to-edge joining is much the same with this machine as with any of the hand-held models; however, with all but the very largest stock, it is far more convenient.

The one joint that this machine cannot make is a framing-type joint located more than 5 inches from either end of the board or on long stock. (See Illus. 57.) These joints must be made with a hand-held joiner.

My only other grievance with the machine is that the dust-spout size, 1¾ inches, is not standard for vacuum cleaner pickup. If you use a Delta 49-255 shop-vac, there is a ready-made coupling; the rest of us have to improvise.

Should these last couple of very minor complaints dissuade one from using the Delta 32-100 stationary joiner? Certainly not. The portable joiner is, of course, a beautiful idea, and for my shop I would buy one of those first; but, as you discover how handy biscuit joining can be, you'll find that this tool can contribute greatly to the efficient operation of your shop.

VII
Elu 3380 Joiner/Groover

The Swiss-made Elu 3380 is a joiner/groover. Not only does it cut slots for biscuit joints, it can also be used for various kinds of trim-saw applications. This 600-watt machine runs at 7,500 RPM and weighs seven pounds. From two feet away, its noise readings are 97 dB (no load speed) and 103 dB (under load). Its cord is 10 feet, 7 inches long.

The Elu joiner has a momentary-contact-only switch, which seems to me to be the safest kind for most portable power tools. Its fitted metal case, with injection-moulded layout liners, is the most compact of the cases available. (See

Illus. 58 (left). The Elu joiner/groover in its straight upright position. Illus. 59 (above). Side view of the Elu joiner/groover.

Illus. 60. Despite its many advantages, the Elu is awkward to use when its mitring attachment is in place.

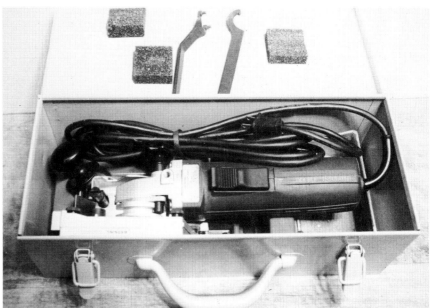

Illus. 61. The Elu in its case. The bottom of this case is a piece of fitted plastic.

Illus. 61.) The manual, however, appears to serve the purpose of liability reduction far better than the purpose of consumer education.

While the joiner has not always been easily available in the United States over the years, it is worth noting that Elu is now a Black & Decker subsidiary, and Black & Decker will use its collective marketing skills to ensure a ready supply of both tools and service.

The Elu joiner is powerful and rattle-free. It features two operating positions, for it works as a groover as well as a biscuit joiner. As a left-handed woodworker, I have to say this model has the best chip ejector. (See Illus. 62.) Unlike all the other joiners that shoot a lapful of chips at me with each cut, this one throws the chips away from me. Unless you are working in an area that must remain as "chip-free" as possible, this chip ejector may be even better than the dust extractors offered as accessories for some of the other joiners.

Illus. 62. Front view of the Elu joiner/ groover. Don't be dismayed by the number of handles and adjustments you see; as you use the machine, you'll note that each one not only serves a purpose, it is also desirable. One advantage of this joiner is that it's the only one that throws sawdust away from the left-handed operator rather than towards him.

The joiner's basic operation is a bit different from that of the other joiners. The manufacturer claims that its primary function is to cut the recesses for joining pieces of wood and/or wood products using glue and flat dowels or biscuits, but its action seems at first to be a bit less precise than that of the other joiners. The operator must pivot it, rather than push it, into the work to get the plunge-cutting effect. To do this, hold the guard/shoe assembly firmly against the workpiece as you pivot the motor assembly. If you are less than perfectly steady or the work slips in its clamp, the well-machined base may move.

However, there is an advantage to using the Elu joiner. For easier access in tight situations, you can pivot the motor assembly with respect to the shoe and guard by loosening the locking knob and rotating the tool to the desired position. Be sure to retighten this knob before proceeding with other adjustments.

To adjust the depth of cut, turn the depth-of-cut adjuster clockwise to cut shallower, and counterclockwise to cut deeper. The depth scale shows the depth the cutter protrudes beyond the bottom of the shoe.

For plate joining, the following sizing information is imperative: biscuit size 0 is equal to 8 on the cutting scale; biscuit size 10 is equal to 10 on the cutting scale; and biscuit size 20 is equal to 12 on the cutting scale.

Side-to-side blade adjustment is not available on the other joiners to the same extent it is on the Elu. (See Illus. 63 and 64.) Of course, all the tools will permit coarse adjustment with a fence, but the Elu additionally offers a fine adjustment by rotation of the handscrew at the rear of the main shoe. Turning the handscrew clockwise moves the blade towards the fence; counterclockwise moves it away from the fence. The manufacturer cautions the operator not to adjust the handscrew while the motor is running, nor to adjust the screw to

Illus. 63 (above) and 64 (below). As you can tell from the photographs, the Elu joiner/groover can be moved laterally nearly a quarter of an inch from the seam where the blade cover meets the base.

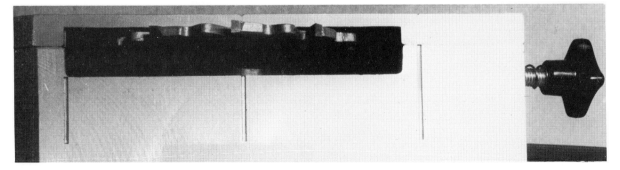

either extreme, as the cutter may hit the guard. I discuss this feature in further depth later on in this chapter.

As with most other joiner operations, try your settings on scrap pieces first. Mark the location for the biscuit recess in both pieces to be joined. It is usually easier to mark the work on the side adjacent to the side that is being slotted. Clamp the workpiece securely. Then, holding the joiner firmly in both hands, align the mark on the guard cover (visible through the sight hole in the side fence) with the mark on the workpiece.

Laying out joints with the Elu joiner is more similar to laying out joints with other joiners than I initially thought. All you have to do to join the pieces is mark the line on the higher of the pieces.

To plunge the cutter into the work, start the motor and pivot the motor assembly. As already mentioned, this joiner differs from the standard joiner because it requires a pivoting rather than plunging motion from its operator. At first this may seem awkward, but after you get use to it, it will not be a problem.

The Elu joiner's blade has 12 teeth, twice as many as most of the blades for the other joiners. It looks like the most substantial of all the blades and I've been tempted to replace the blade on my other joiners with a blade like this. (See Illus. 65.)

The Elu joiner has a plastic anti-splintering insert, as shown in Illus. 66, which should be inserted when you cut into veneered panels. Assemble the insert into the grooves in the guard, and cut a slot through the insert by placing the tool against a piece of soft wood and pivoting the cutter down through the insert and into the wood. The only problem I can see with this otherwise excellent accessory is that apparently only one piece comes with the tool, and the manual does not mention how to order additional pieces. I anticipate needing many more of these pieces.

Another advantage of the Elu joiner is the ease with which its blade can be changed or adjusted. As shown in Illus. 67, this is the only unit with a blade guard that doesn't have to be removed with a tool. Of course, the blade has to be removed with tools. A C spanner fits between the cutter and guard to the left side of the spindle and hook spanner into the notch on the inner clamp washer; because the threads are "left-handed," you have to turn them clockwise to remove the blade. Be sure to check the direction of the cutter rotation when replacing the blade.

Because the Elu joiner angles into the cut rather than plunging straight in, this unit sometimes jitters when it is being used. Clamping the work is imperative.

This joiner and the Bosch joiner are the only units tested that lack spring-loaded positioning pins or other means of holding the work while joining, and unless the grooving feature is important to you, this may be a major failing of the machine. On the other hand, if you could learn to overcome this short-coming, it should be noted that the general design and construction of the machine are the best among the joiners.

As shown in Illus. 68–70, when the Elu is disassembled it seems to have more in common with the other joiners than originally thought. It is similar to a 4-inch grinder, especially on the inside. However, unlike the other machines, the blade on the Elu joiner adjusts precisely to the machine, or to about $\frac{3}{16}$ inches to either side of it. This feature when coupled with the

Illus. 65. The blade of the Elu joiner/ groover has twice as many teeth as most of the blades on the other joiners.

Illus. 66. This plastic insert fits into the mouth of the Elu joiner/groover to prevent chipping when grooving or slotting in fine veneers.

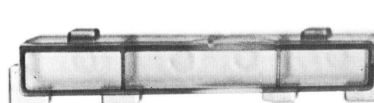

Illus. 67 (left). The Elu joiner/groover with its faceplate removed. Illus. 68 (above). Even though the Elu joiner at first looks very different from other joiners, when you disassemble it you'll note that it looks very similar to a 4-inch grinder—as do most of the other joiners.

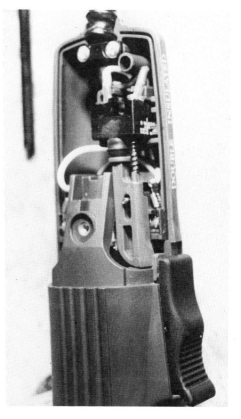

Illus. 69 (above). An interior look at the Elu joiner/groover showing the main gear. Illus. 70 (right). A partial look at the inside workings of the Elu joiner. Note the motor compartment.

Illus. 71. Back view of the Elu joiner/groover with the rip fence in place.

fence, as shown in Illus. 71, makes the unit handy for grooving. It also allows for fine positioning of the slots when slotting.

To fully appreciate this, you might have to experience what I discovered as I built a small compartmentalized box with a Lamello Top. After the edge joints were cut and fitted, I started preparing the interior pieces for mounting. Here the trick is to line up the same edge of each joint, cut the vertical slots first, and then, after sweeping away the chips, cut the horizontal slots. The marking out is very simple—a few small marks are all you need to line up your work.

I discovered that the Lamello Top's blade is too high for centered cuts in ½-inch stock; a standard joiner won't make this cut without a lot of manipulation. This is where a machine like the Elu comes in especially handy. I fastened a half-inch-piece of Baltic birch plywood with biscuits on four sides; this plywood was ½ inch thick, and the other pieces in the project were ⅝ inch thick.

I positioned the Elu joiner's cutter exactly in the center of the half-inch plywood, as shown in Illus. 73, and cut a groove rather than just individual biscuit holes. This turned what could have been a most unpleasant repair into an easy job. (Remember when using the Elu joiner or any other joiner that it is the operator, not the tool, who makes costly mistakes. Therefore, always measure twice, and cut once, and be sure that you have the machine's settings secured tightly in place.)

Once when I was using the joiner I noticed that there were too many chips on the floor and decided to start cleaning up. The broom I was using hit the cord of the Elu unit too hard. The Elu unit hit the floor—a fall of about three feet. There was absolutely no damage to the joiner. Kudos to the design engineers at Elu who understand that we are now and again likely to drop a tool. Nevertheless, I hesitate to repeat this unintended experiment with the other models.

One improvement that I would love to see is a cordless version of this or any other joiner. A joiner that sits in a continuous charger while it's not in use would be ideal. The manufacturer who offers a high-quality version of this type of joiner will have at least one ready customer: me.

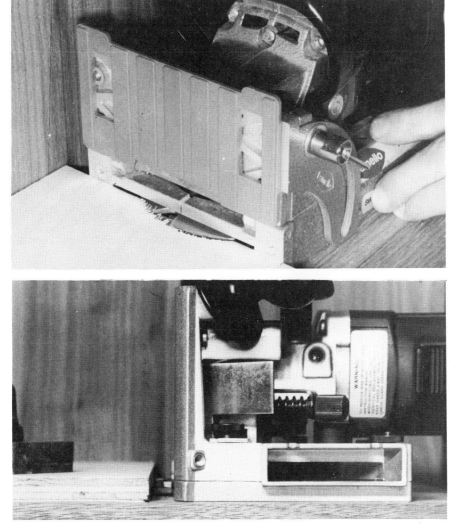

Illus. 72. When the Lamello Top is used to make a centered cut in a piece that will be carcass-mounted, the operator will have difficulties in laying out the joint.

Illus. 73. The Elu joiner/groover with its adjustable blade height can easily center a cut in a piece that will be carcass-mounted.

VIII
Freud JS-100 Plate Joiner

In shops where it will see more extensive and difficult use than in my furniture shop, the Freud JS-100 will prove to be a favorite, especially at its very low price. (See Illus. 74.) This six-pound, three-ounce machine has a 550-watt motor that produces 10,000 RPM, a 7-foot, five-inch cord and an outstanding injection-moulded case that closely resembles an attaché case. (See Illus. 75.)

Illus. 74. Side view of the Freud JS-100 joiner.

Though several joiner users I know think that the Freud machine is crude, I don't agree with them. Freud is one of the newest entrants in the market, and the manufacturer has done its market research very well, for the unit contains most of the best features of the other models, and it will do almost anything they can do. It would be foolish to expect all the features in the Freud model to be as sophisticated as those on joiners that are much more expensive. After all, the Freud JS-100 is just two-thirds the price of the next least-expensive unit.

Freud's switch is well positioned, but it doesn't slide as easily as any of the others; it appears to be located so that it can be operated with the thumb on the operator's right hand. Right-handed operators will also like its dust ejection, but left-handers may want to try the machine before they buy it, as there is no dust-collection attachment yet available for it. It's faceplate, shown in Illus. 76, is scaled in both inches and millimetres, and the 45-degree or 90-degree depth adjustment plate is grooved for a square operation and fitted with the easiest-to-adjust lock-down knobs found on any fixed-angle faceplate joiner (Illus. 77); the setup could actually be done one-handed if necessary.

The manufacturer cautions users to change the brushes every 250 hours, and to have a service center change every third

Illus. 76 (above left). Front view of the Freud JS-100 joiner with its 45- or 90-degree depth-adjustment plate removed. Note that the faceplate is scaled in both inches and millimetres. Illus. 77 (above right). Freud JS-100 joiner with its 45- or 90-degree depth adjustment plate in place.

brush to ensure a thorough cleaning, etc. When I first read these instructions, I thought that Freud is either doing a better job of telling joiner users what to look out for or it may be anticipating problems that will occur when the joiner is being used. After discovering that the manual for the Lamello Top joiner makes the very same suggestions, I realized that Freud is just stressing prudent maintenance precautions.

Since this is the first fixed-angle machine to be discussed, let's examine a few setup techniques that are unique to this type of machine. With the tool unplugged, set up the depth-of-cut with the two knurled nuts on the right side of the machine. After you have done this the first time, you are ready to cut all three size biscuits, and you will only have to check the setting occasionally. I set the depth-of-cut with a marked size 20 biscuit rather than a ruler. To mark the biscuit, simply draw a line from tip to tip, and then set the machine by setting the blade slightly wider than the mark.

You will save a lot of time if you have a set of gauge blocks in the most common thicknesses of the wood you use. Loosen the adjustable faceplate and set it on the gauge with the base of the joiner flat on the bench. Using these gauges is better than relying on the tool's "square" slides. (See Illus. 78.)

Illus. 78. An assortment of depth-adjusting gauges of varying thicknesses.

The slots on the mating pieces must be exactly parallel. While there is some play lengthwise, there is absolutely no play in the width of layouts. This set of blocks will also make it easier to set up the tool for stacking biscuits where you need extra mechanical strength in the joint, particularly when joining thicker stock.

On the faceplate, there are two positioning pins that are parallel with the center of the blade. These pins help to keep the JS-100 from moving as you make your plunge cuts. If you use the unit for slotting rather than plunging, you can easily remove these pins by unscrewing the faceplate. The maximum-depth-of-slotting cut is just over half an inch.

The Freud and Porter-Cable joiners certainly provide the most value for the money. If I had never used the Lamello Top, I would have been more than easily satisfied with either of these machines. The very popularity of the Freud JS-100 may make it hard to get. Calls to dealers in various parts of the United States got the same response: "Sorry, out of stock!"

IX
Virutex O-81 Joiner

The Spanish-made Virutex O-81, shown in Illus. 79 and 80, is imported by Holz Machinery/Rudolf Bass, Inc. Its 500-watt motor produces 10,000 RPM. The machines weighs seven pounds and has a 7-foot, 3-inch cord. From two feet away, its noise readings are 101 dB with no load and 104 dB loaded. A dust collector, shown in Illus. 81, is available as an option. Its full-year warranty appears to be the industry standard.

In many ways, the Virutex is more like an inexpensive Lamello Top than like an upscale Freud joiner. In fact, the Lamello Junior joiner was introduced as direct competition for the market the Virutex tapped into. This machine and the Lamello machines have been on the United States market the longest. The Virutex appears to be used in more American

Illus. 79 (above left). A top-down view of the Virutex joiner.
Illus. 80 (above right). A side view of the Virutex joiner.

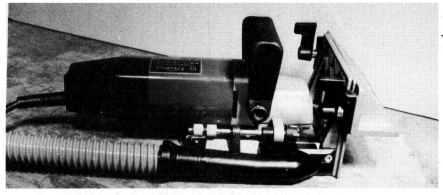

production shops than any other joiner with the possible exception of the Lamello Top.

The Virutex's manual, while far from comprehensive, is arguably the best of the manuals that accompany the lower-priced machines; this is another factor that should help the tool find its way into more and more amateur shops.

The switch on the Virutex joiner is superior to that on the Freud model because it is easier to operate and is well-isolated from the user under a dustproof, "shock-proof" rubber gasket. There is also a disadvantage to using the switch: It is rather awkwardly placed on the top near the rear of the machine. One cannot assume that correct stance for operating the joiner and then turn it on. It remains to be seen as to whether the position of the switch will lead to any more problems.

The castings of the faceplates and other parts of this machine are of heavier-gauge aluminum and have more ribbing and more accurate layout lines than the castings on the Freud and the other less-expensive models. The faceplate on the Virutex is ruled in both inches and millimetres on each side of the casting. Another desirable feature is its quick-adjust depth-of-cut scale that permits the user to select number 10, 0, or 20 biscuits in virtually no time at all and switch easily from size to size within a given application.

Though the Virutex joiner seems longer than the other joiners, it isn't; however, there is no chance of pinching fingers while operating it. Indeed, it is one of the most comfortable joiners to operate, and it becomes even more comfortable with its optional dust-collection hose in place. A vacuum-assisted dust-collection system is more of a necessity than optional, especially if you use the tool left-handed.

Illus. 82. A front view of the Virutex joiner reveals the inch and metric scales on each side of the faceplate; this is a great aid when you are setting up the cutting gauge and other attachments.

The fence adjusters are accurate and square, but you can't turn the large lock-down knobs, shown in Illus. 83, through a full 360 degrees without hitting the rear supports of the faceplate. When the screw that holds the tightening knob in place fell off one side of a Virutex I was using, I discovered that it was easier to adjust the machine, and became convinced that I could use the machine for a long time without replacing that screw. If the machine had been my own, I would have promptly removed the other side's mating screw.

The Virutex is one of two machines whose case doesn't strain the "relief" portion of the cord. (The "relief" portion to the cord is the first 4–6 inches of the cord as it comes out of the body of the tool. This portion is quite a bit thicker than the rest of the cord, and provides a "strain relief" against the possibility of the cord being twisted until it breaks or is pulled out of the tool.) Furthermore, the "used" Virutex I sampled demonstrated the longevity of these injection-moulded cases; the case, as shown in Illus. 84, appears to have been used a great deal, yet it remains serviceable. Though it shows signs of wear, it shows no signs of deterioration.

Since the Freud joiner has appeared on the market, the price of the Virutex has come down considerably for the careful shopper. It is easy to understand how some joiner users I know

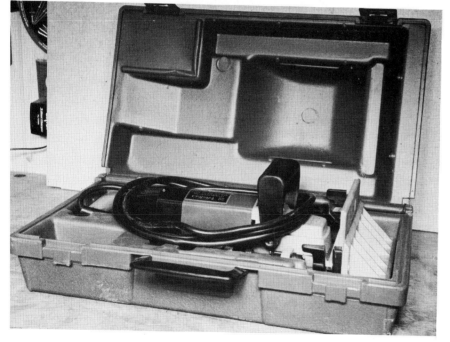

Illus. 84. The Virutex joiner in its durable case.

preferred this machine over all the others. Price played a factor in their choice, for this machine is more like the Lamello Top than any of the other less-expensive models. In fact, at approx-imately less than half the price of the Lamello Top, the Virutex might be the machine of choice for someone who really doesn't need the Lamello accessories and the Top's flap-front faceplate. Its quality is such that it deserves wide and continued sales.

X
Lamello Junior and Standard

The Swiss-made Lamello Junior is a 500-watt, 10,000-RPM machine that weighs less than six and a half pounds. Its cord is 8 feet, 2 inches long. From two feet away, the noise level is 101 dB with no load, and 102 dB in operation.

The Junior manual is barely adequate. However, Colonial Saw provides instructions for various kinds of operations that almost make the manual unnecessary.

When I assembled a test panel for a joiner article in *Popular Woodworking*, our unanimous least favorite joiner was the Lamello Junior. At that time, it seemed to be an inferior version of the Top, and several things were troublesome. For instance, the Junior is the only joiner in which you have to add clips to the machine's depth-of-cut gauge to change biscuit sizes; the clips (Illus. 85) aren't hard to add, but they are certain to get lost, especially in a woodworking shop.

Another problem was noted when I handed the machine to a long-time professional woodworker for his appraisal. He plugged it in and was ready to use it, but he couldn't find the switch. This toggle switch (Illus. 86) works easily, but is located inside a recess on the unit's back end, which is frustrating.

The faceplate (Illus. 87) is plastic, as is the fixed-angle fence (Illus. 88). And though some plastics are superior to cast aluminum in resistance to wear, etc., this was not the type of plastic used. Also, the scale is marked only in millimetres.

Finally, as Illus. 89 shows, the body of the Junior is shorter than that of the other joiners (the body of the tool is 6⅛ inches, compared to the Top's 7⅛-inch body) to the extent that when we tried to operate it by holding it by the body rather than by the handle—which is the way the plate joiner is operated—our fingers were pinched.

To some extent, each of the above problems still exist. On the other hand, there are a number of reasons to consider

Illus. 85 (above). The Lamello Junior with one of its spacing pins in place. Illus. 86 (right). The switch on the Lamello Junior is unhandy because of where it is positioned.

Illus. 87. The plastic faceplate on the Lamello Junior is marked only in millimetres.

Illus. 88. The Lamello Junior with its fixed-angled fence in place.

Illus. 89. This rear shot may show better than any other the essential differences between the Lamello Top (right) and the Junior (left). The dust-ejection kit shown on the Top can also be used on the Junior.

the Lamello Junior. Most shops use size-20 biscuits in far greater quantity than either of the other sizes—indeed, almost to the exclusion of the other sizes—so the spring clips really shouldn't be a problem. And, Lamello does package the clips in a handy plastic vial, so if the operator is careful, he won't lose them.

The machine operates smoothly and, amazingly enough, is the quietest of the die-grinder-type joiners, which in my shop is very important. It will operate all of the accessories in the Lamello system. (Accessories for patching pitch spots and for making two different kinds of knock-down joints are discussed in Chapter XIV, as are other accessories.) If you find the accessories attractive, this compatibility is a considerable bonus.

Additionally, the Junior is the only "moderately priced" unit equipped with a slip clutch in its mechanism. This is a big plus if the people who operate the tool in your shop are not totally familiar with its operation.

In an attempt to keep costs down, Lamello only provides a cardboard case with the Junior. (See Illus. 90.) Actually, this cardboard case is one of the better cases, and doesn't harm the "relief" portion of the cord.

Illus. 90. The Lamello Junior in its carton. Despite the carton's cardboard construction, it is one of the most desirable cases available.

When working on the joiner article for *Popular Woodworking*, I concluded with this paragraph concerning the test results: "As I read through these comments about the Junior, I find that I am describing a machine that is too good to be 'worst.' Maybe what made it 'worst' is that it came to me in a carton with the 'best' (the Lamello Top). If Lamello moved the switch and replaced some of the plastic with metal, it would be easy to move this one far up the scale."

In fact, Lamello has gone a long way to make the Junior more attractive. It can now be retrofitted with the Lamello Top tilting front plate. (See Illus. 91 and 92.) The operation takes less than ten minutes and can be done with just the tools supplied with the Junior. The following parts are needed: front-plate with flap and two screws; bracket with two screws; and a clamping lever. Additionally, two more accessories are strongly recommended. They are the right-angle guide with screws and the thickness plate (4 mm).

Neither Lamello nor Colonial Saw will make Junior machines that have this modification; nor will they accept parts removed from the Junior for "credit."

Making this modification will give you a machine with nearly all the quality and function of the Lamello Top for considerably less than you might expect to pay for the Top. You will save even more if you don't need either the right-angle guide, which is used for installing Lamello Paumelle hinges or working with stock over ¾ inch thick, or the thickness plate, used for working with ½-inch stock and for some mitring applications. For these tasks, return the Junior's standard faceplates. It takes just a little while to change the faceplates, and the savings may be worthwhile.

With this modification, the Lamello Junior becomes an attractive choice. It can even be used as a second joiner for someone who wants to trim face frames, as used with the Lamello Nova, or patch bad areas when used with the Lamello Mini-Spot. All things considered, my *Popular Woodworking* review seriously underrated Lamello's Junior joiner.

Newly available in North America is the Lamello Standard, which is, in fact, the Junior motor with the Top faceplate; it is priced about midway between the Top and Junior machines, and offers the advantages of both.

Illus. 91 and 92. The Lamello Junior being used with a Lamello Top tilting front plate.

XI
Kaiser Mini 3D

The Kaiser Mini 3D is available only from W. S. Jenks & Son, 1933 Montana Avenue, N.E., Washington, DC 20002. This 600-watt, 5.7-amp, 10,000-RPM tool, which weighs just under six and a half pounds, appears to be exquisitely made. Its cord is almost ten feet (3 metres) long.

The Kaiser castings are absolutely first-class, almost as fine as the castings on the Elu and Lamello machines for machining accuracy. They are too well made for them to be covered as they are with bright orange paint! The high-quality components in the motor housing are laid out very logically. (See Illus. 94.)

Though there are not many repair centers, you may not require much service for this finely crafted unit. However, it is

Illus. 93. The Kaiser joiner with its two fixed-angle face-plates.

73

Illus. 94. The layout of the motor for the Kaiser joiner.

not known at this time if W. S. Jenks and Son will be able to provide quality service for the joiner. The company claims to have a complete parts inventory and to be willing to cannibalize stock machines for parts if necessary. However, this has yet to be proven. This uncertainty about the warranty service may be a real drawback to an otherwise finely crafted unit.

As I write, the Kaiser joiner is a very new entry in the United States market, and an English-language manual isn't yet available. The metal case appears to be logically designed, as Illus. 95 shows, and built to last a lifetime, but the plywood insert on the very first Kaiser joiner case that I saw was broken. Since the joiner I tested was neither brand new nor supplied by the importer, it is hard to determine whether this is a design flaw, the result of some abuse, or a combination of the two. The bottom of the case has one-inch-diameter felt feet to prevent damage to the furniture when the tool is being used.

The joiner is quiet, reading 98 dB from two feet away at no load, and 96 dB as the tool plunges into the work. In fact, it is quieter than all the other joiners with the exception of the Porter-Cable 555.

Notably absent from this plunge-type joiner is the handle that seems so unnecessary on all the other similar joiners. Instead, this joiner features a handle which comes straight out five inches on either of the tool's sides. The left-handed user will like this handle much better than all the other handles, especially when it is mounted on the "right" side.

Another notable difference is that while all the other plunge-type joiners have a pair of spring-loaded positioning pins that help to stabilize the machine while in operation once it has been positioned, this one has a ribbed, non-slip rubber faceplate. Tests have demonstrated that this surface grabs at least as well as the spring-loaded pins.

The spring that returns the joiner to its "ready" position is so strong that only one is needed; it is held in place by a very small shear pin.

The Kaiser joiner features a switch that works very much like the switch on the Elu joiner, but this one can be locked in position. Like the Lamello Top, the Kaiser joiner has a spindle

Illus. 95. The Kaiser joiner in its well-laid-out case.

Illus. 96. The underside of the Kaiser joiner with its combination blade guard/faceplate removed.

lock so that the blade can be removed with only one wrench. Unlike all but the Bosch and Elu joiners, access to the blade on the Kaiser machine, shown in Illus. 96, is as quick and easy as loosening two screws; in other words, this is the only plunge-type joiner with a truly easy-access blade.

This tool has both an anodized aluminum fixed-angle faceplate for both straight and mitre cuts at 45 degrees and a flap faceplate that works like, but does not look like, the flap faceplate on the Lamello Top joiner. (See Illus. 97 and 98.) Unlike the Top's flap faceplate, this one is not the full width of the joiner, but only 3⅜ inches of the 4¾-inch total width of the face. Furthermore, it lacks the 4-mm plate that permits the Lamello Top to work so effectively in thin stock; this absence is made conspicuous by the fact that the flap faceplate is ²⁵⁄₃₂ inch from the base, a great distance. The fence that holds the flap at various angles is somewhat small, but it works. It seems to be designed so that the flap can cut various angle joints from the outside of the work.

The Kaiser joiner compensates for its shortcomings in several ways. The fixed-angle faceplate fits very "tightly" and is square. The knurled nuts that hold it in place can be finger-tightened, but if you want an even tighter fit, they can be turned a bit further with a coin. The height-adjustment scale, shown in Illus. 99, measures very accurately to 50 millimetres (almost

Illus. 97. The Kaiser joiner with its fixed-angled faceplate in place.

Illus. 98. The Kaiser joiner with its flap-front faceplate fixed in a 90-degree position.

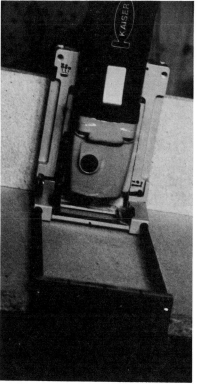

Illus. 99 (above). Profile of the Kaiser joiner. Note the mounting slots for the fixed-angle faceplate, and the height adjustment scale. Illus. 100 (right). The long fence on the Kaiser joiner.

two inches), but it measures only in millimetres. The grooves that hold the faceplate also accept a fence with 10¼-inch guide bars. (See Illus. 100.) Using this anodized fence, one could lay out joints with the Kaiser machine that would be very difficult to lay out with other machines.

Also, the Kaiser machine has some real advantages when it is being used for grooving. With the other joiners, one must lay out a straightedge to lay the machine against while grooving. The fence permits one to position grooves relative to the edge of the work.

At the right rear of a base that appears somewhat bulky (I say appears because the measurements aren't noticeably greater; perhaps it looks bulky because the handle is missing) is the depth-of-cut setting gauge, which will permit cuts in size-0, 10, and 20 slots. (See Illus. 101.) The fine adjustment is as difficult to make as the coarse adjustment is easy. Of course, the depth of cut isn't affected when the cover is removed, so the greater precision this affords may be a real advantage.

Illus. 101. The depth-of-cut gauge on the Kaiser joiner. Note the removable three-position cam and the difficult-to-get-to fine adjustment.

While I question the Kaiser joiner's long-term potential, it is a machine that has a lot of features and which can be bought at a relatively favorable price. While there is presently no information about accessories, and while the joiner will not run any of the components of the Lamello system, it appears to offer most of the features of the Lamello Top for little more than the price of an upgraded Lamello Junior. If the importer were to make the optional dust extractor with its four-metre hose a standard feature, the Kaiser could be an excellent choice.

The chip-ejection system is very efficient, but a recent 80-biscuit project that I made with it made such a mess that I became convinced that a joiner should not be used without a dust-collection system.

XII
Lamello Top

The Swiss-made Lamello Top is likely to be at or near the top of everyone's list of desirable joiners. Smooth of operation, and extremely convenient, the Top is perhaps the fastest joiner and the easiest one to use. Its 600-watt, 8,000-RPM motor cuts without hesitation in even the toughest material. This seven-pound machine has a cord that's 8 feet, 3 inches long. Its wooden case, shown in Illus. 102, is excellent, although probably an inch too short for proper maintenance of the cord's strain relief. From

Illus. 102. The Lamello Top in its case. The case is slightly too small for the cord.

two feet away, the sound level is 100 dB unloaded and 102 dB under load.

Even the Top's manual is excellent, the only joiner manual that can be so rated. The clarity of the text and photos is exceptional. Indeed, the only not-so-desirable aspect to the Lamello Top is the fact that it is the most expensive of the joiners.

Rather than the fixed-angle faceplate for mitring and butt-joining, the machine has an adjustable faceplate that can be set for any angle. Illus. 103 shows the faceplate set at 50 degrees, an angle that would be much more difficult to cut with fixed faceplate joiners. Because of the faceplate's adjustable head, the operator does not have to make jigs and supports to join angles other than those at 90 or 45 degrees. There is both a mechanism for adjusting the angle of the flap faceplate and a 4-mm attachment plate that can be used to center the joiner on both ¾- and ½-inch material. This can be done by using the attachment plate to change the position of the blade relative to the flap-front faceplate. This plate also helps speed the spacing of multiple-width biscuit joints in thick material.

This 4-mm snap-on auxiliary faceplate is useful for working joints in thin stock. It serves as a marking gauge for laying out internal and corner carcass cuts as well as an aid to actually cutting them. Lay the auxiliary plate on the line to be cut, and scribe the short line indicated by the arrow on the faceplate as shown in Illus. 106 and 107; once at either end of the joint should be enough. Then lay the piece perpendicular

Illus. 103. The Lamello Top set for 50 degrees, a setting that would be much more difficult to get on fixed-angle joiners.

Illus. 104. The Lamello Top set for 90-degree cutting; the plastic shim for cutting half-inch material is in place.

Illus. 105. The other side of the setup shown in Illus. 104. The tabs that mount the shim for half-inch material are clearly visible, as is the adjuster for the depth of cut.

Illus. 106 (left). The 4-mm accessory plate used with the Lamello Top. Illus. 107 (above). The plate is used to mark out interior carcass joints in thin stock.

Illus. 108 (above left). Making the vertical cuts on the joint laid out with the 4-mm accessory plate. Illus. 109 (above right). To make the horizontal cut on the joint laid out with the 4-mm accessory plate, you have to lay out the piece to be cut over the edge of some material and cut it with the accessory plate in place.

Illus. 110. The Lamello Top with its variable-thickness right-angle guide.

against these newly scribed lines the same way you would lay it against the regular layout line, clamp it in place, and cut your vertical slots, as shown in Illus. 108. Then, with the auxiliary faceplate in place, cut the horizontal slots; make sure the surface to be cut is over the edge, as shown in Illus. 109, rather than still clamped in place; this way, the auxiliary faceplate will correctly center the cuts.

This joiner makes repetitive cuts more accurately and

more quickly than the other joiners, and from countless hours of hands-on use with this and other joiners, I'm convinced that there is less vibration transferred to the operator's hands.

While I hoped for finer, more precise indexing marks (which aren't really needed, since the slots are slightly longer than the biscuits), for more gradations marked on the adjustable-angle flap-front, and for a scale on the faceplate (even an all-metric scale would be preferable to no scale—especially when the machine's fixed-angle faceplate is being used), nearly everything else about the machine is better than one might expect or need. Its powerful motor is slip-clutch-protected, its switch is the easiest to use of the joiner switches (though I still prefer the switch on the Elu joiner), and the machine has arguably the best balance. The Lamello Top has a spindle lock that makes blade changing markedly easier. Indeed, every feature you've come to appreciate on the other joiners has probably been modelled after the features on the Lamello Top.

Lamello seems intent upon designing the system around this tool; any new accessories they offer will run from the Top. To date, these accessories include a dust extractor, a glue bottle, hinges, a centering awl, the Nova edge trimmer, the Mini-Spot patching kit, Lamex knock-down fittings, Simplex knock-down fittings, installation kits for both kinds of knock-down fittings, and a set of clamps. These are described in depth in Chapter XIII.

If you buy a Lamello Top, one of your first projects may well be a new case; the supplied case is nicely made and lovely to look at, but it is about an inch too short to house the cord without straining its "relief" cord. You'll find yourself making all kinds of projects with the joiner just for the simple joy of using a machine that is so finely conceived and designed.

XIII
Lamello Top-10

The Swiss-made Lamello Top-10, which has superseded the Top, is almost certain to appear at or near the top of everyone's list of desirable joiners. The name Top-10 is a reference to ten features on the joiner that are an improvement on the original top's features.

Smoother and more convenient than the Top, the Top-10 is perhaps the fastest and easiest joiner to use that's available today, and arguably the best-balanced. There is also less vibration transferred to the operator's hands. Its 700-watt, 10,000-RPM motor cuts without hesitation in even the toughest material.

Illus. 111. The Lamello Top-10 (left) has superseded the Lamello Top.

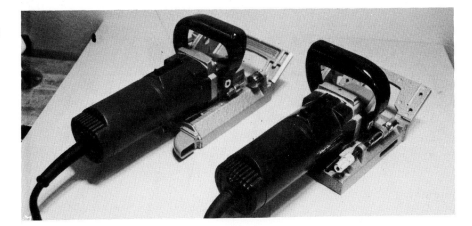

Illus. 112. Side view of the Top-10 set up to cut mitres from the outside of the joint.

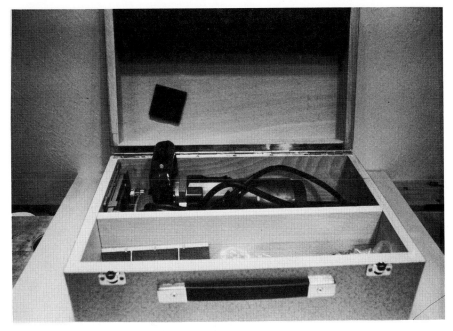

Illus. 113. The Lamello Top-10 in its case.

The machine weighs seven pounds and has a cord that's 8 feet, 6 inches long. Its wooden case is much larger than that of its predecessor—one no longer has to worry about damaging the strain relief cord—and it has a very convenient accessory storage space.

The sound level of the joiner is 93 dB loaded or unloaded. This is nearly 10 dB quieter than the Lamello Top. The machine comes from the factory with a pair of dust outlets, one of which can be attached to a piece of vacuum cleaner hose of your choice rather than the expensive attachment that is supplied with the Top.

The joiner retains the Top's flap-front faceplate, which now has an accurate protractor added to it. Its fixed-angle attachment is a modification of the Top's; it has sliding dovetails

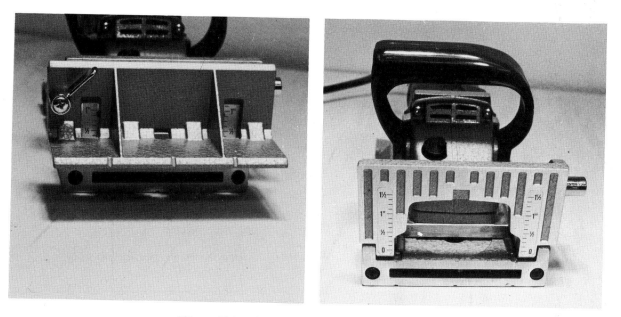

Illus. 114 (above left). The Top-10 with its fixed-angle attachment showing read-through scales. Illus. 115 (above right). The Top-10 with its flap-front faceplate.

for quick and accurate attachment to, and removal from, the flap-front faceplate. There is now a height scale that can be read through the back of the attachment when it is used. (See Illus. 114 and 115.)

The joiner's adjustable head reduces the need for making jigs considerably; the Top-10 doesn't require us to make jigs and supports for joining angles other than 90 or 45 degrees.

The Top-10 retains the 4-millimetre attachment plate that permits the joiner to center on both three-quarter and half-inch material; this plastic plate changes the position of the blade relative to the flap-front faceplate. The plate also helps speed the spacing of multiple-width biscuit joints in thick material.

The powerful motor on the Lamello Top-10 is slip-clutch-protected, and its switch is the easiest one to use (though I still prefer the switch on the Elu joiner). It has a spindle lock that makes blade-changing markedly easy, and an easy-off bottom for blade-changing or for changing dust-ejection chutes. It has a pair of anti-slip pads instead of the spring-loaded positioning pins. They are just as effective.

The Top-10 has six depths of cut on its depth scale: 0, 10, 20, max, D, and S. (See Illus. 116.) The first three represent the

Illus. 116. A closeup view of the depth-of-cut mechanism on the Lamello Top-10.

stock Lamello biscuit sizes. Max indicates maximum depth of cut, S the depth for Simplex connectors, and D the proper depths for Lamello Paumelle hinges.

The only aspect of the Top-10 that isn't a great improvement over the Top is the manual. The Top's manual is excellent, still the only joiner manual that can be so rated.

The least-desirable aspect of the Lamello Top-10 joiner is its high list price. However, it is such a well-made machine that if you could afford the luxury of the added expense, perhaps you should consider buying it.

Any accessories Lamello offers can be used with the Top-10. These accessories are discussed in depth in the following chapter.

XIV
Commercial
Accessories

Ever anxious to make a good product better and to profit from after-market sales, the various joiner manufacturers have introduced a variety of accessories. Lamello is the major manufacturer of accessories for plate-joining. It even sells an entire joining system. Most of the accessories discussed here are, in fact, components of the Lamello system. Some of these accessories are genuine innovations, and others appear to be devices designed to profit from the typical (American?) craftsperson's desire to acquire gadgets. Remember, all the accessories discussed here, with the possible exception of the Nova and the Mini-Spot, will more or less work as well with a non-Lamello plate joiner. Following is a survey of the accessory market that will hopefully separate those useful devices from the useless ones.

Lamello Gluer

The Lamello gluer, shown in Illus. 117, is the handiest accessory for plate-joining. Though it is expensive, it is extremely useful. It is made of plastic and has a wooden base. You can store the glue bottle upside down so that the glue is always near the tip of the bottle and is ready to be used instantly all the time. The metal tip releases the glue through a pair of channels (Illus. 118) on opposing sides, thus delivering the glue to the sides of the cut slots. This is a definite advantage

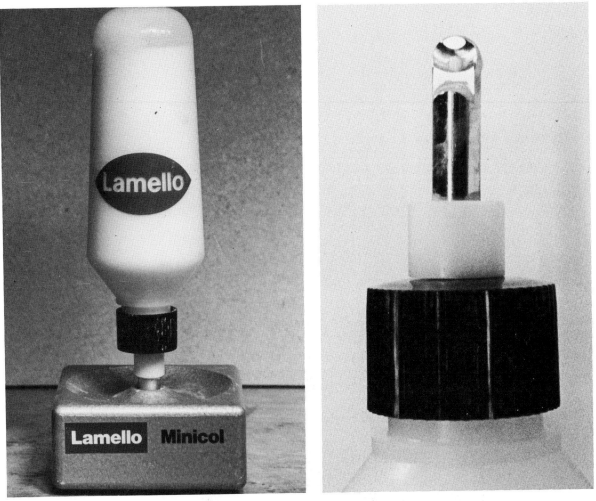

Illus. 117 (above left). If you want to get full value for your joiner, the Lamello glue bottle is an almost indispensable product; it saves almost as much time in the workshop as the joiner itself. Illus. 118 (above right). A closeup look at the tip on the Lamello glue bottle.

because for many kinds of joints the only areas that need glue are the insides of the biscuit slots, and the glue bottle makes the work neater and quicker. The joiner and this gluer made it possible for me to do profitable renovation work that I would have been unable to try in my pre-joiner days. The gluer is indispensable to anyone who has to work with prefinished materials.

No matter which plate joiner you buy, you should use the Lamello gluer. It or something very much like it should be standard equipment with every brand of joiner. It simplifies all biscuit-joined assembly, makes the assembly neater and faster, and will save a great deal of glue at assembly time. Your next bottle of glue will probably last twice as long.

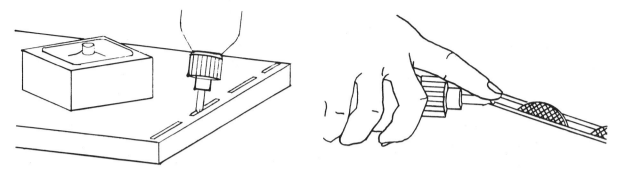

Illus. 119–122 (above and below). The Lamello glue bottle is useful for gluing biscuit slots (above left), the slotted surfaces themselves (above right), slots for unusual biscuit configurations (below left), and for internal gluing (below right).
(Drawings courtesy of Colonial Saw)

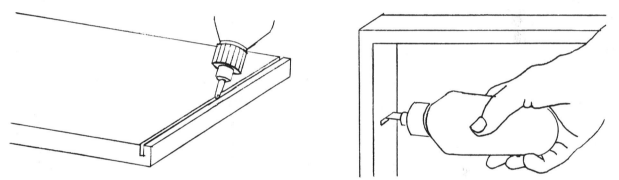

Dust-Collection System

Lamello, Virutex, Kaiser, and Bosch each have vacuum hoses on the market for collecting the dust thrown by their joiners; while I haven't figured out the reason why these dust-collection systems are so expensive, it doesn't take long to figure out their usefulness. With a dust-collection system in place, the joiner runs so cleanly that you can run it in the living room, if necessary. These dust pick-up units take a little time to install, though, so after the installation, most people "plug" the power cord into the receptacle and the dust-hose attachment into the vacuum cleaner. Lamello's dust-collection system, at 134 inches, is almost a yard longer than its Virutex counterpart (Illus. 123) and its hose is smoother on the inside. The best way to run the unit with the dust pick-up accessory is to attach

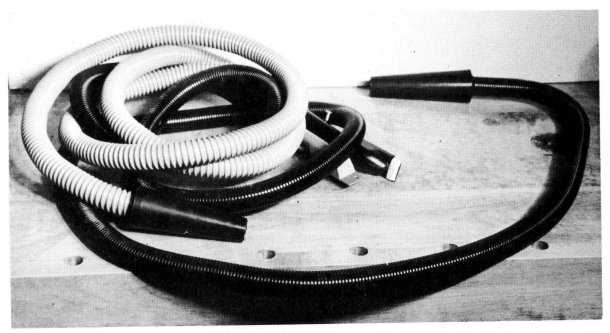

Illus. 123. The Lamello dust-collection kit, the darker of the pair shown here, is nearly a yard longer than the Virutex kit shown with it.

Illus. 124. This left-handed operator really wishes there were a dust collector attached to the Lamello Junior he is using.

the hose to the power cord with narrow pieces of duct tape, and then operate the tool with the cord and the vacuum hose both thrown over your shoulder. The dust pick-up kit is necessary, especially if you are left-handed. (See Illus. 124.) The availability of this accessory should be a prime consideration when choosing a joiner, even if you elect not to purchase the pick-up right away.

When I first started surveying joiners, I could hardly wait for the dust collector to arrive for my Lamello Top. Once it arrived, however, I debated whether it was better to live with the dust or cope with the 10 feet of 1¼-inch hose protruding from the machine all the time. Once I discovered that many kinds of wood dust are carcinogenic, I kept the dust collector on the joiner.

Lamello Hinge-Mounting System

Lamello's hinge-mounting system has to be one of the best solutions I've encountered to the problem of hinge-mortising. Lamello Paumelle hinges come in three types: bright nickel, black, and solid brass. Each hinge will hold 22 pounds, so it may be necessary to use more than a single pair in some operations. The hinges are sold as "left" and "right," so they can be used in combinations for nonremovable doors, or separately for doors that can be lifted out of position. (See Illus. 125.)

These high-fashion hinges, regarded by some as the "hallmark of contemporary furniture," can be quickly installed. After you have sawn your box open and smoothed and squared the opening, clamp the door to the body. Mark out hinge locations with a pair of simple pencil scribes. Position the plate joiner so that it is centered exactly on the seam. If you are using the Lamello Top joiner, I am almost certain that you will have to

Illus. 125. Lamello Paumelle hinges are strong, attractive, and easy to mount. The pair shown here is for both left- and right-hand mounting. You can use mixed hinged types for doors that are to be attached in place. Since each pair of hinges can hold about 20 pounds, more than a single pair may be required for large doors.

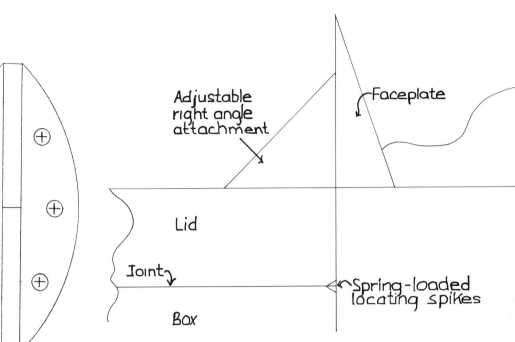

Adjustable
right angle
attachment

Faceplate

Lid

Joint

Box

Spring-loaded
locating spikes

Illus. 126 (left). Measuring for depth of cut against the Lamello Paumelle hinge. Illus. 127 (above). Cutting the slot to mount the Lamello Paumelle hinge. Note that the spring-loaded positioning pins are centered on the seam of the opening. Use every available stabilizer. Careful work is essential here.

Illus. 128. Lamello Paumelle hinges can be used in various applications. A: They can be used on folding partitions and doors. Lamello hinges offer easy removal of doors for cleaning and moving. B: They can be used on drop leaves and doors; rotate one hinge for fast, permanent installation. C: The hinges are perfect for box lids and full overlay cabinets. D: Lamello hinges can be used on flush cabinet doors for European styling.

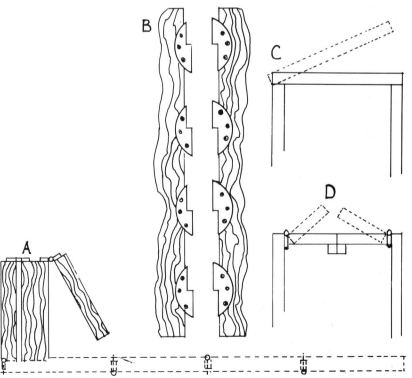

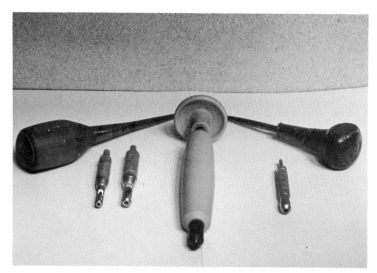

Illus. 129 (left). The Lamello awl is made specifically to install Lamello Paumelle hinges. Illus. 130 (above). The Lamello awl with a Stanley centering awl, a pair of standard scratch awls, and a pair of Vix-Bits.

use the fixed-angle faceplate rather than the flip-front. Cut kerfs the size of number 20 biscuits for each hinge. Unclamp and separate the door from the carcass. Set the hinges loosely in place. Punch through the screw holes with an awl. (See Illus. 129.) Screw the hinges in place.

Lamello's awl is surely handier than the hammer-like version that I have used for the past many years. Despite its high price, a regular user of Lamello Paumelle hinges may find this awl to be superior to other available awls or to the drill-centering bits (sold sometimes as Vix-Bits). (See Illus. 130.)

Lamello Lamex Knockdown Fittings

Lamello recommends the Lamello Lamex knockdown fittings for panel, leaf assemblies, and shelves. The plastic inserts come in white or brown, as shown in Illus. 131. The wooden biscuits, size 20 only, are fibre-reinforced. (See Illus. 132.) An installation kit, shown in Illus. 133, makes things easier by allowing rapid and accurate cutting of the T-slot for the connecting part. The kit is not a necessity, but I would not care to install these fittings without it. It makes their installation accu-

Illus. 131 (above left). The Lamello Lamex knockdown fittings. This system includes brown and white plastic inserts with adjustable metal clamps (shown disassembled at top), and fibre-reinforced size-20 biscuits. Illus. 132 (above right). This Lamex biscuit has been severed with a knife, and only after great effort, to reveal the fibres.

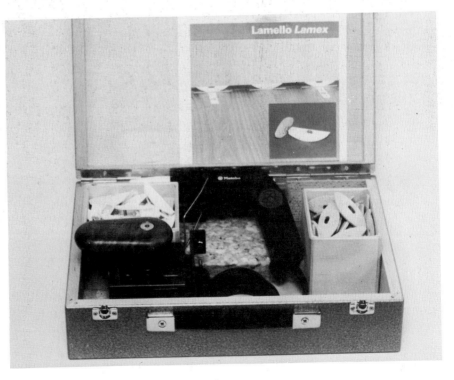

Illus. 133. The Lamello Lamex installation kit offers a complete method for joining units that must be frequently disassembled.

rate and quick; while installing Lamex fittings freehand is possible, it will take too much time and will not be accurate for most jobs.

To install the Lamex system, align the pieces to be joined as in a regular biscuit-joined operation; cut the mating slots, and glue the special wooden biscuits in place, as shown

Illus. 134. Here are the "biscuit" ends of the Lamex knock-down clamping system.

Illus. 135. Close-up of the Lamex cutter gauge that shows its relationship to the biscuit slot, to which it is cutting a wide perpendicular slot.

in Illus. 134. Place the Lamex cutter gauge, shown in Illus. 135, in a slot on the other side, and make three or four plunge cuts, all the way through the "swing" of the gauge. (See Illus. 136.) Set the Lamex connectors into these milled cuts, and glue them into place with the hot glue gun, as shown in Illus. 137 (I imagine that in this case, a neat epoxy job could be done in-

Illus. 136. The Lamex gauge at work.

Illus. 137. Gluing the Lamex connectors in place with a hot glue gun.

stead). After all the adhesives have set, insert the biscuits into their mating holes and tighten the screws. (See Illus. 138.) The result is a tight, removable (and thus portable) joint, a joint whose tightness can be adjusted. This may well be the finest system that I have encountered for making knockdown furniture.

As I have said of too many of these accessories, Lamex fittings are expensive. So is the installation kit, which consists of a cutting gauge, a glue gun, a case, and some samples.

Lamello Spanner Set

The Lamello spanner set can be bought in a variety of ways. You can buy the entire set, which consists of two clamps that have eight metres of belting each, two tension hooks, and four 24-inch aluminum profile corner protectors. (See Illus. 139.) If all you want is the spanner itself, you can buy one pair of clamps that have five metres of belting. Clamps can also be bought without belting, and the nylon belting can be bought separately. So can the tension hooks (per pair), the 24-inch profile corner protectors (per set of four), and the 5-inch profile corner protectors (per set of four). (See Illus. 140.) As expensive as it is, the Lamello spanner set is one of the most useful clamping systems available on the market, and it is a great enhancement to any joining system.

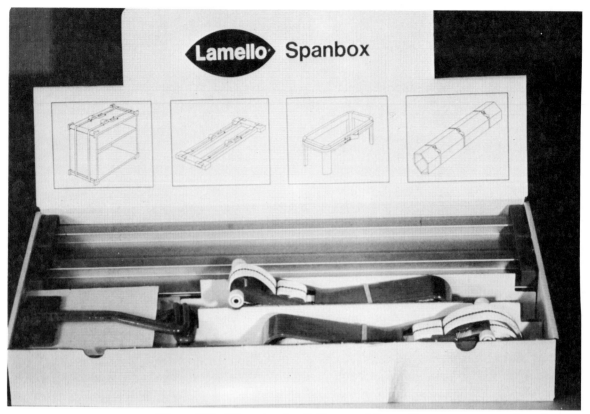

Illus. 139. Here is the full set of Lamello spanners, a band-clamping system that is so simple and ingenious that many woodworkers will wonder how they got along without it. This system is especially useful for joining carcass units with mitred corners.

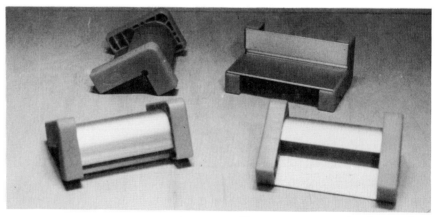

Illus. 140. The five-inch corner protector for the Lamello spanner set.

The advantage of the Lamello spanner set is that a lone woodworker can quickly glue up carcasses that might otherwise be impossible to glue up when he is working alone. It also helps ensure that the joints are drawn tight. For carcass work, the Lamello spanner set is much handier than using bar or pipe clamps, and it may leave the carcass light enough to be moved if necessary. To use it in carcass construction, simply set the "bottom" joint in the extrusion with sufficient webbing to span the unit hanging out proportionally. Assemble the carcass. Set the remaining pair of extrusions on the "top" joints, and "set" the clamp lightly.

After you have made the final adjustments for square, tighten the clamps. (See Illus. 141 and 142.)

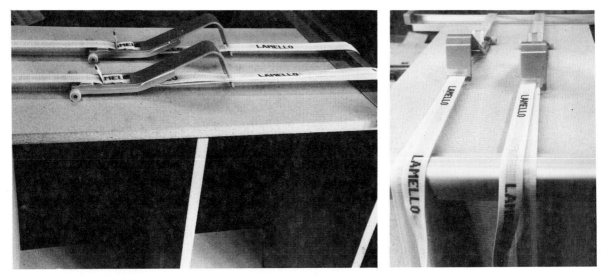

Illus. 141 (above left). The Lamello spanner at work. Illus. 142 (above right). The Lamello spanner shown over the 24-inch corner protectors.

At first, the spanner set seems less handy for flat-clamping. However, when a standard pipe or bar-clamp is tightened, you can move the work slightly by torsioning it clockwise. The spanner set does a better job of maintaining flat, true work, particularly in thicker stock.

Those are the advantages to using the spanner set. The following point may be considered a drawback by some. Glue squeezeout and other drippings are not considered a problem with standard clamps. We expect to get glue on them. When it does get on them, we let it dry and then chip it off. In this respect, our gluing habits will have to change. The webbing of the Lamello spanner kit won't last long with glue spilled all over it. If you are using the spanner with biscuit-joined assemblies, there should be little if any seepage.

Here is a clamping system that may replace eight or more pipe clamps in carcass construction and is immeasurably quicker than other clamping systems. If time means money in your shop, as it does in mine, and if you expect to always get first-class results, the Lamello spanner set may have a place in your shop. If, however, the price remains an obstacle, consider having your woodworker's guild or club buy a set for the occasional use of all the members.

Lamello Mini-Spot

The Lamello Mini-Spot, shown in Illus. 143, is a tool that has very little value in the typical American cabinet shop. I am so accustomed to treating in a different way the problem that the Lamello Mini-Spot addresses that I found this attachment to be expensive far beyond its value to me. Basically, the Mini-Spot consists of a spear-point saw blade and a special faceplate for

Illus. 143. The Lamello Mini-Spot patching kit attached in place on a joiner.

Illus. 144. A Mini-Spot patch made of pine. These patches are also available in maple, ash, cherry, beech, spruce, and other species.

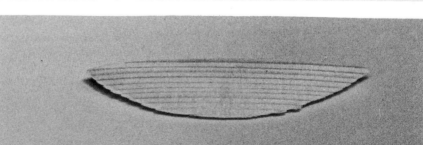

the Lamello Top or Junior joiners. After the plate joiner has been converted, simply run it over pitch pockets or similar defects, and then glue in patches that Lamello sells separately in various wood species. (See Illus. 144.) After the glue has set, pare the patch flush to the surface.

This procedure is as involving as the one I use to solve the same problem. I cut out my stock in four- to six-inch widths for gluing, using the defects as markings for cutting lines. The Mini-Spot would make it possible to use wider boards if you had a project where wider boards would be genuinely desirable, and if you could keep those wide boards from warping.

Part of my objection to the tool is that one must take a few minutes to set up the machine for the Mini-Spot application, an important consideration in shops where the price of labor is almost always higher than the price of materials. But even this is secondary to the visibility of the patch, which at best looks no better to me than a shop-cut circular plug.

Lamello Nova Edge Trimmer

The Lamello Nova edge trimmer shown in Illus. 145 and 146, is an attachment that is probably of most use to kitchen installers. You can attach it to any Lamello machine quickly with a pair of heavy screws; an adjusting screw permits the blade to contact the cabinet flush and makes it possible to trim a face frame flush with the edge of the cabinet. However, it doesn't seem to be sufficiently superior or quicker than trimming with

Illus. 145. A view of the Lamello Top with the Nova attachment in place.

Illus. 146. The Lamello Nova edge trimmer can be quickly and easily fitted to your Lamello Top or Junior for trimming protruding edges with great precision.

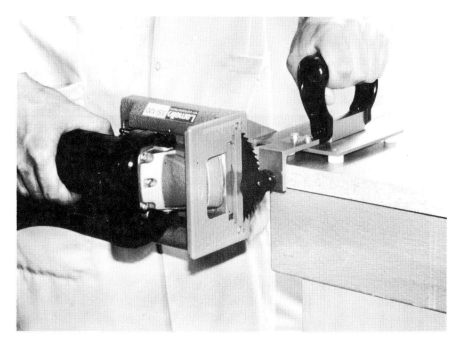

a flush-sided dovetail saw and a strip of sandpaper, as I have always done in the past. And the flush-cutting dovetail saw is just a fraction of the price of the Nova. If I were a professional kitchen installer, I might want a Nova permanently installed on a Junior; this way, the savings in time would eventually pay for the tool.

The Bosch and Elu joiner/groovers seem to be easily better than the Nova for performing this operation. The Bosch machine will do edge-trimming to a depth of $\frac{7}{8}$ inch with the biscuit-slotting blade or with a very fine 22-tooth (2.2-mm) blade. The combination of Bosch's small size, handy rip fence, generally well-designed body, and fine blades make this machine much handier than the Lamello Nova, much less a standard portable circular saw, for this sort of edge-trimming. The Elu's lateral blade adjustment permits it to complete this task with amazing accuracy.

Lamello Inserta

If you are going to use your joiner in a production shop, you may be interested in the new Lamello Inserta (type A-85), shown in Illus. 147–149. This low-maintenance pneumatic machine holds 50 biscuits, and the biscuits can be hand-loaded or purchased in a special Inserta pack for quick loading. An ingenious quick-release mechanism for loading permits the spring tension to be held off the plates while freeing the locator arm for loading the biscuits. The manual is as good as the one for the Lamello Top, but using the accessory is so easy it is almost superfluous. Colonial Saw calls the market for this tool "somewhat limited," and its high list price is undoubtedly a contributing factor to that limited marketplace.

Illus. 147. The Lamello Inserta pneumatic biscuit installer.

Illus. 148 (above left). This front view of the Lamello Inserta shows the prongs that guarantee dead-center insertion every time. Illus. 149 (above right). Close-up of the "business" end of the Lamello Inserta.

Lamello Stationary Attachment

Lamello has just released an attachment for mounting Top and Junior joiners to stationary columns; it even has its own ratchet version of this attachment. (See Illus. 150 and 151.) Lamello recommends these attachments especially for use with the Lamello Mini-Spot attachment, but they might also be useful for joining small pieces as well.

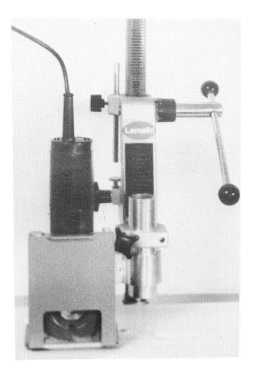

Illus. 150 (above). The Lamello stationary attachment can be used for mounting to any stationary column. Illus. 151 (right). The Lamello stationary attachment mounted to press apparatus.

Maintenance and Cutting Techniques

XV
Caring for Your Joiner: Maintenance and Troubleshooting

Many of us woodworkers are incredibly lax about periodic maintenance of our tools and equipment. Following a regular maintenance schedule is not just a question of keeping expensive tools working, but also a question of personal safety. Properly lubricated, sharp tools are safer than dull ones, for they don't require the operator to force them.

Unless you are certain that you can put it back together without making a mistake, don't take your joiner apart. It is surely less expensive to pay an authorized service center to do routine maintenance than to have to pay them to fix a tool you've reassembled improperly. Even after having taken dozens of tools apart, I am still apprehensive when I first operate a tool I have disassembled and reassembled. There are many things that can go wrong.

Check your tool's warranty. If the warranty is still good, return the tool to your dealer or factory-authorized service center rather than opening it yourself. Attempting to repair the tool on your own, even opening its outer assembly, will almost certainly void the warranty.

If you have to disassemble the tool, pay particular attention to how the parts fit together. The tool's parts list is not sufficient to ensure proper reassembly. Here are the basic steps to disassembling the tool:

1. Lay out the parts in the order that you take them off the tool.
2. Remember how "tightly" each part is fastened so that you can return all the parts to the same torque. Where possible, use an appropriate torque wrench for this work; if, like myself, you don't have one, be sure you have a good memory.
3. Most of the fasteners on plate joiners are metric. Handle them with care, as they are harder to replace than inch-size fasteners. Use metric nut drivers and wrenches where needed; if you strip the shoulders off the hexagonal nuts, you will make it much more likely that next time you will have great trouble with disassembly.

Certain repairs shouldn't even be attempted at home, but you should at least know how to change the blade and check the motor brushes. All the manufacturers specify that the second change of brushes (third set of brushes) should be done at a factory-authorized service center along with thorough cleaning and, generally, replacement of all gear grease. Let the trained technicians at these centers disassemble the tool when the brushes have to be changed.

To change the blade, do the following: First, remove the blade's cover. On the Bosch, Elu, and Kaiser units this is a simple matter of removing a couple of screws or knurled nuts. On the other plunge-type joiners, you must remove the springs with the hook provided (Illus. 152 and 153) and the knurled

Illus. 152. Removing the spring on the Virutex machine. Note the large hold-down clamps on the fixed-angle fence.

Illus. 153. Removing the pair of springs is an early step in disassembling the joiner.

nuts that adjust the depth of cut. (See Illus. 154.) After they are off, remove the faceplate from the slide assembly with a Phillips screwdriver. (See Illus. 155 and 156.) Be sure to keep the stabilizer pins and their pressure springs in a safe place.

At this point, the rest of the cover just slides off, leaving you with an exposed blade, as shown in Illus. 157. Remove the

Illus. 154. Removing the jamb nuts that adjust the depth of cut is another early step in disassembly.

Illus. 155 and 156. To remove the faceplate, you have to remove this pair of screws.

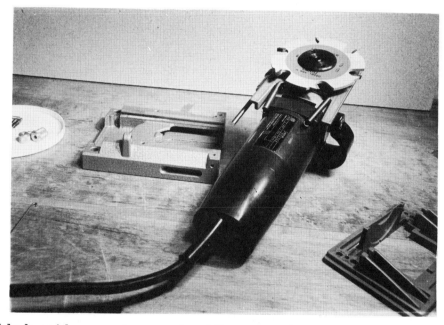

Illus. 157. A Lamello Junior with its faceplate, depth-of-cut nuts, and sliding base removed. This amount of disassembly only takes a moment and is necessary when making most equipment changes and when periodically cleaning and maintaining the machine.

blade with wrenches or the Allen wrench and spanner provided; these steps are shown in Illus. 158 and 159. On the units with a spindle lock, you'll only need one wrench.

On each machine, the blade is removed by turning it counterclockwise. While the blade should require sharpening only approximately every several boxes of biscuits, this disassembly should be done fairly regularly for other reasons, like cleaning, oiling and other fine-tuning.

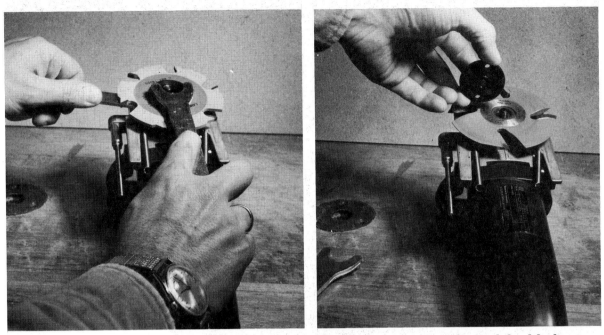

Illus. 158 (above left). Two wrenches are required for blade removal on all the joiners with the exception of the Lamello Top (which has a spindle lock). Illus. 159 (above right). The last step in removing the blade is unscrewing this arbor.

Cleaning is an equally important part of the process. Be sure to clean dust and debris from the inside of the tool as you reassemble it. The unit should be blown out regularly with compressed air. (See Illus. 160–162.) Be sure to wear safety glasses while cleaning with compressed air. From the slide assembly remove all traces of the residue formed by the mixing of sawdust and lubricating oil. Use toothpicks to remove small specks of this residue that are likely to accumulate in the corners. As you replace the slide assembly on the tool, lightly lubricate it. (See Illus. 163.) This lubrication should be lightly repeated every few days or whenever the slide is not working as smoothly as possible, but not so often that the oil clutters up the work surfaces.

Motor brushes on the machines are all changed the same way. All manufacturers claim that the tool will simply stop before the tool's armature is destroyed by the carbon brush's spring. Ideally, this is so, but if the joiner is not maintained properly, this may not happen. Therefore, approximately every 50 hours of use, remove the motor cover, generally held in

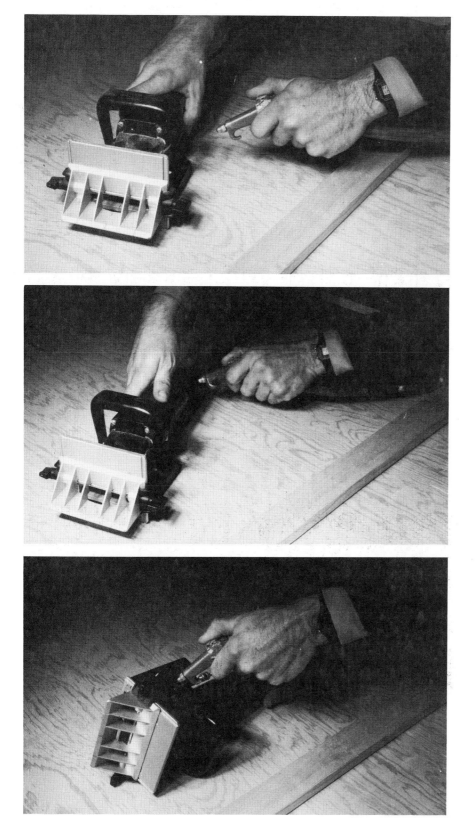

Illus. 160–162 (left and below). The joiner should be blown out regularly with compressed air in the areas shown.

Illus. 161.

Illus. 162.

Illus. 163. Apply oil regularly but sparingly to both sides of the joiner's track.

place by just one or two screws, and inspect the brushes. This should not take more than a minute, for the brush holders on all machines are easily accessible once the cover is off. Pay attention to the manufacturer's warnings and replace the brushes only with identical replacements; to ensure optimum motor life, replace them in pairs. (See Illus. 164 and 165.)

While you have the motor cover off, it is a good idea to blow the accumulated dust out of the motor. If you don't have a compressor, a can of "photographer's air," which does not cost much, would make a wise investment. Another advantage to photographer's air is that there is never any liquid in it; liquid sometimes accumulates in compressed air lines.

Changing the belt on the Porter-Cable 555 blade is a unique procedure. After you have removed the blade, remove the four screws on the base plate; this gives you access to the belt and pulley. Be sure to replace the blade with the exact-toothed "super torque" belt.

Following are some of the problems you may have to deal with when using a joiner:

1. Cord damage is almost inevitable in most shops. Replacing the cord shouldn't be difficult; remove the cover and attach

Illus. 164 (above left). Sometimes you have to use a pair of needlenose pliers to completely remove tight motor brushes. Illus. 165 (above right). Shown is the joiner motor. Note the partially removed motor brush.

an identical cord in the same manner as the damaged cord was installed.

2. A clogged dust chute is invariably the fault of the operator. After you stop the machine and unplug it, clean the dust chute by probing it with the unit's spring removal hook. Also, never run a unit with a chip extractor without also running the vacuum cleaner; it only takes a few chips to jam the hose, and cleaning it will be a major challenge.

3. If you find that your unit is "skittish" when you cut very hard material, check your blade for sharpness and review your operating methods. Perhaps you should hold the tool more firmly.

4. If you have dropped the tool, check the slide assembly carefully to ascertain that it hasn't been bent out of shape.

Attending regularly to proper maintenance and storage is extremely important. Clean, sharp tools are safe, economical tools.

XVI
Edge-to-Edge Joining

While the butt joint is the joint most commonly associated with the joiner, let's begin by looking at edge-joining, because edge-joining generally has to be done before the panels are ready to be butt-joined. It takes only about an extra half a minute per joint to cut slots and insert biscuits when edge-joining with biscuits, and the results will be a joint that will hold tightly without sliding around that will be flat and true even before you plane or sand it.

Illus. 166 shows the "poor" side of an edge-joined walnut panel on which no sanding, scraping, or planing has followed assembly; the only preparation this panel received was that the very narrow glue bead was removed before it had hardened.

Illus. 166. Note the marked seams in this edge-joined walnut panel. No sanding, scraping, or planing has followed assembly.

The arrows on the panel mark the seams. Very light sanding will make this panel as near perfect as is possible. The "snipe" at the top of the panel will be cut off. As you can tell from Illus. 166, with a joiner it is quite possible to get along with a small but accurate surface planer.

To ensure the flattest possible work, glue only two pieces together at a time. Using this method, follow these easy steps for efficient edge-to-edge gluing as when making panels: Start with opposite sides of a wide panel, and glue two or three pairs of pieces at a time until you have one pair left, and then glue these pieces together. The layout is very easy. Mark the boards to be joined 2 inches from either end and about 8–10 inches apart between them. (Illus. 167 shows a rather closer placement, but 8–10 inches is almost ideal for edge-joining.)

Illus. 167. Marking out a piece for edge-joining. I made the marks that appear on the piece more prominent than usual because I wanted them to show up in the photograph.

Edge-joining is most easily done by placing the fixed-angle cutting guide or the flap fence flat at 90 degrees over the edge, as shown in Illus. 168, and cutting the edges and ends.

When edge-gluing panels, insert the biscuits about 10 inches apart. This interval is an estimate that is based on practical experience. In my shop, I tried three different biscuit placement options when gluing panels for a commercial project. On one pair of boards I placed a biscuit approximately every eight inches; many biscuits were used and this seemed to be a waste of material. On the next pair of boards, I placed biscuits every 12 inches; those biscuits were spaced so far

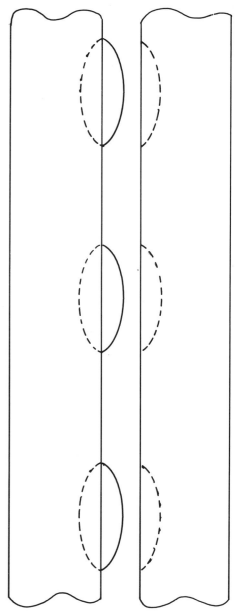

Illus. 168 (above). Cutting the slots for edge-joining in a small board. Boards this small don't really need to be biscuited together unless you have to use the wider piece immediately. Illus. 169 (right). When edge-joining, remember that the biscuits should be 8–10 inches apart.

apart that the joined materials "wandered" and were uneven. Biscuits clamped at ten-inch intervals turned out to be just about right.

Biscuit-joining like that used for edge-to-edge joining not only ensures flatter panels, it also allows you to remove clamps much sooner than you normally would. Since it becomes much harder to run out of clamps, your workshop will function much more smoothly.

XVII
Butt Joints

Corner Butt Joints

If joining joints quickly is important to you, you'll be intrigued by what you'll read here. Making standard corner butt joints is much quicker than dovetailing or dowelling.

There are two ways to make the standard corner butt joint used on internal members like partitions, dividers, etc. Both these methods will be described here. Decide upon one after trying both and use it, but not to the exclusion of the other. Each method has its advantages.

In both methods, accurate setup is the key to accurate work. The joint is laid out exactly the same. (See Illus. 171.)

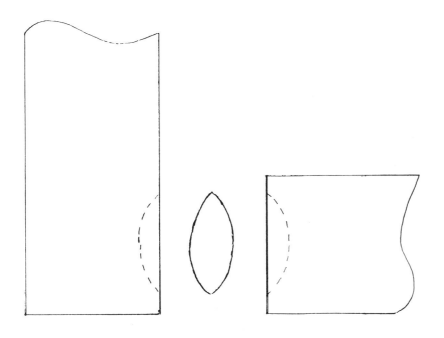

Illus. 170. Butt corner joint used for framing work.

Illus. 171. Marked-out carcass butt joint.

Mark lines two inches in from each end of the joint and four to six inches apart between them. Make parallel rows (one or more from each side) if the workpieces are 1 inch thick or thicker. Layout is fast; the layout lines are only used to ensure that your slots are within a quarter inch of one another throughout the length of the joint. After you have worked with the machine for approximately an hour, you'll be able to mark out your joints quite easily by eye rather than with a scale or template.

Next, cut the joints. In the first method, cut a slot on the face of one of the pair of pieces to be joined and cut a slot on the edge of the other piece. (See Illus. 172 and 173.) This method puts the slots at an equal distance from the outside edge on both boards, which is exactly what is needed. The drawback to this method, however, is that the joint can get out of square if you don't cut accurately into the sides, which isn't that easy to do.

In the second method, lay out a piece of material adjacent to your work that's the same thickness as that being joined. This material supports the machine as it works through the joinery. Stand the pieces edge to edge, perpendicular to one another; carefully lay out the vertical edge piece over on its axis, which is its inside edge. Scribe a line at that inside edge to ensure accurate positioning. Clamp the vertical piece to the

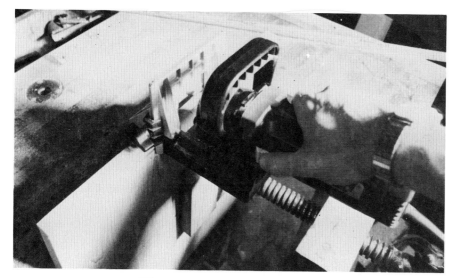

Illus. 172. The over-the-edge method of cutting produces results that are less accurate than those achieved through the second method.

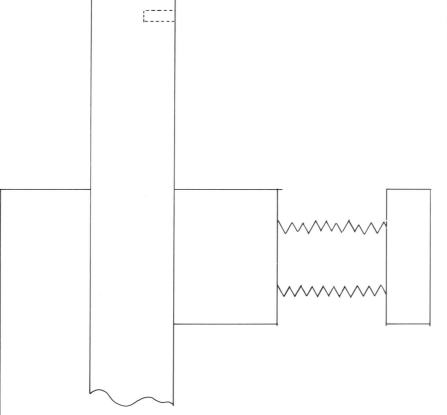

Illus. 173. In the over-the-edge method of cutting, the joiner rests on the narrow top edge of the board to be joined.

horizontal piece so that the surfaces to be joined are at right angles to one another. Mark out the joints, again 2 inches from either end and about 4 inches apart in between. Mark out the joints only at the very edge of the "top" piece, as shown in Illus. 174. Using the extra piece to help support the joiner squarely (Illus. 175), first cut the vertical slots, as shown in Illus. 176 and 177. Then, after sweeping away the chips, cut the horizontal slots, as shown in Illus. 178 and 179.

Whichever method you choose to cut the joints, they glue up alike. After unclamping the pieces, put glue in the slots, insert the biscuits, and assemble. There is no need to glue the end grain to the long grain, so it will pay handsome dividends if you do some preliminary finishing work before assembly. Glue as follows: Lay on its side one of the pieces that will carry the insert. Glue only the slots; either run a fair bead of glue down each side of the slot or use a plate-joining glue bottle. Insert the biscuits in each slot. Next, glue the slots on

Illus. 174. Shown here are the layout for a carcass corner butt joint and the shim piece (at bottom) that will keep all the cuts square.

Illus. 175. The joiner rests squarely on the left block, C, while you cut first A, and then B.

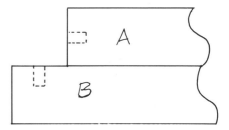

Illus. 176 and 177. Cutting the vertical slots with the shim piece in position. This method is slightly more difficult but rather more accurate than the over-the-edge method of cutting shown in Illus. 172 and 173.

Illus. 178 and 179 (next page). Two views that show how to cut the horizontal slots with the shim piece in position.

Illus. 179.

the pieces to be attached; then attach them immediately. Apply the next batch of glue and biscuits and finish the assembly.

Proceed in this fashion for the entire project. Make sure that the assembly is physically possible according to the way you're proceeding. Test-fit the piece with dry biscuits before gluing it. Keep an extra 1,000 biscuits on hand. Even a very small project will take more biscuits than you have planned for.

Standard Butt Joints

Making standard butt joints to join internal carcass members is only marginally different from making corner butt joints.

When laying out an interior joint, as on a drawer frame, shelf, etc., lay out the joint to one side of the member rather than to its center. (See Illus. 181.) In other words, if you want to center a ¾-inch piece exactly, lay out ⅜ inch to the side from which you plan to work; this will produce a centered joint. The pieces must be laid out logically: Since you will be marking the sides (top or bottom, front or back) of the joints, use the same side all the time. Be sure to label where the pieces go; you will forget the assembly order, and that will lead to trouble, especially if you aren't cutting your biscuit slots exactly in the center of the work. As in the preferred method for the corner butt joint, first cut the vertical slots, as shown in Illus. 182 and 183, and then the horizontal slots, as shown in Illus. 184 and 185.

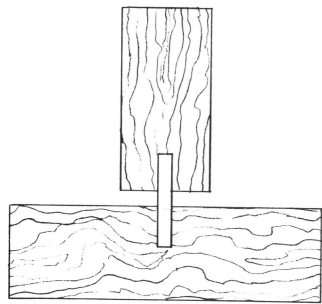

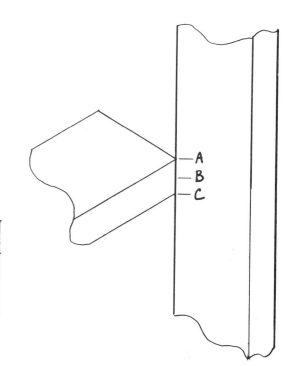

Illus. 180 (above). Edge view of an internal carcass butt joint. Illus. 181 (right). Lay out the interior butt joints at A or C, not at B.

Illus. 182 and 183. Cutting vertical slots for a carcass unit.

*Illus. 184 and 185
(below). Cutting the
horizontal slots for a
carcass member.*

Illus. 185.

Making a small box is a simple procedure that integrates the steps described above. When using plate-joinery to make the box, make sure that the planning and cutting are precise, which is the type of work that should be expected in a production shop.

Laying out the joints is the same procedure already described: Mark two inches from either end of the joint and about

4 inches apart between them. (See Illus. 186.) After you have cut the slots, apply glue and biscuits to them, and assemble the joints as quickly as possible. Illus. 187 shows a carcass joint that was put together so quickly it had to be hastily disassembled for the photograph, which points out how well glue is distributed within the joint. Indeed, if clamping a joint should prove awkward, it can be held together by hand for about ten minutes. It took me less than ten minutes to join, glue, and assemble a small box from accurately cut pieces.

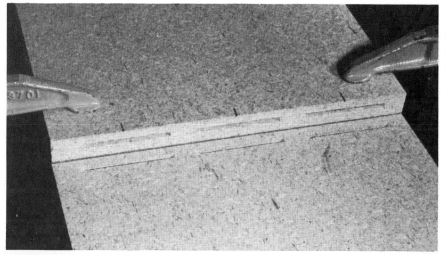

Illus. 186. The horizontal and vertical slots needed for this carcass joint.

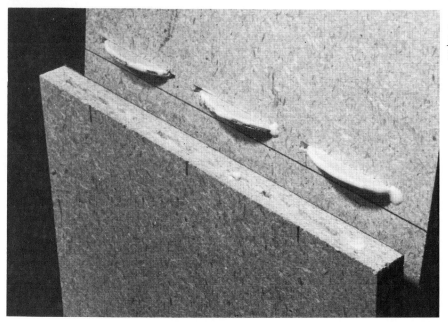

Illus. 187. A carcass joint with the biscuits in place.

Illus. 188 and 189 show a drawer ring being slotted for mounting in a carcass unit. (Also see Illus. 190 and 191.) Chapter XIX discusses how to make this kind of joint in thin stock. To make this joint in stock of normal thickness, again mark two inches from the left, four inches on center in between. If there are many pieces, be sure to label them. Cut the vertical slots first, and then the horizontal ones.

Illus. 188. Cutting mounting slots in a drawer ring. Note that part of the biscuit in the joints that hold the drawer ring together shows, and will have to be chiselled or sawn out before the assembly is completed.

Illus. 189. Another view of the same cut.

Illus. 190 (above). Even though this small carcass unit has a lot of biscuits and slots, cutting and assembly were far quicker than they would have been if another method of joinery had been used. Illus. 191 (right). This seems to be a very awkward way to clamp carcass units.

Apply the glue and assemble. All the pieces that go between two other pieces must be added at once, and you must assemble the joint from the inside out. Failing to assemble all members that were cut from the same part of the logical sequence will mean you will have to omit those parts; they can't be added in between biscuit-joined work.

It would be impossible to overemphasize the importance of partial finishing before beginning biscuit joinery, especially when joining these internal members. At the very least, sand with all but the finest grits of sandpaper. Assembly has to be done so quickly that failure to do the prefinishing will make some of the finishing virtually impossible to do—or at least painstakingly difficult.

After the sanding is done, assemble the pieces without doing any joining; the fit must be almost (if not absolutely) perfect. Correct any imperfections in the fit at this point. Then your finished product will be worthy of your signature.

These precautions also apply when you're making a large project. The hardest thing about making carcass units in a small shop is cutting the material accurately to both square and dimension. If you're working with expensive panel stock, it may be worthwhile to have your lumberyard do as much of the cutting as they reasonably can be expected to do—if they can do the cutting accurately.

Illus. 192 shows that even an extremely wide extension on your crosscutting gauge isn't enough for you to accurately crosscut or mitre wide stock. Even though the extension shown in the photographs is nearly three feet long, it is not much better than the naked mitre gauge, which is not of much use. But if the extension has just been cut, as shown in Illus. 193, it can make a great layout aid if you can remember to tip it so that it is parallel to the saw's table as you measure over the edge with it. A better way to lay out the joint might be to add the thickness of the material you're cutting to the width from the inside edge of the blade to the edge of the table, and then do your layout with a scale, a square, and a straightedge.

Instead of cutting with the crosscutting gauge in place, remove it (and the fence as well) and cut the piece while using

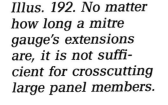

Illus. 192. No matter how long a mitre gauge's extensions are, it is not sufficient for crosscutting large panel members.

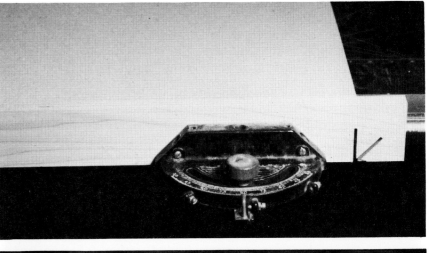

Illus. 193. A freshly cut mitre gauge extension can be a valuable layout tool.

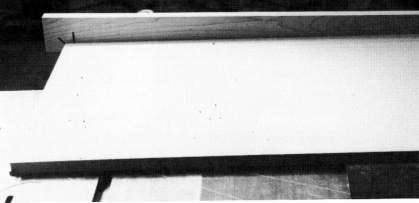

a 2 × 2-inch batten clamped to it as a fence against the side of the saw's table. (See Illus. 194.) The batten should be at least a foot longer than the piece being cut to ensure most accurate "fencing" jobs. There should also be a piece on top of the panel to keep the clamps from marring your work. Check with a square, as shown in Illus. 195, to ensure that you are working accurately.

Illus. 194. A pair of battens that will permit the side of the saw to serve as a fence is better than a mitre gauge.

Illus. 195. Make sure your battens are square before you cut the large members.

The project shown in Illus. 196 and 197 was designed specifically to store some seldom-used writing supplies. Though the finished chest was to be kept in the basement, there were several requirements it had to meet: It couldn't take long to build (more than two hours was out of the question); it couldn't consume more than one sheet of material; it had to be inexpensive to make; and it had to be about the height of a table. All these criteria were met. This project was assembled with butt joints except for the top two joints, which are mitre joints.

Following on page 134 is a cutting list for this project. Techniques for adding a face frame will be discussed in Chapter 8.

Illus. 196. A view of the carcass unit in its clamps.

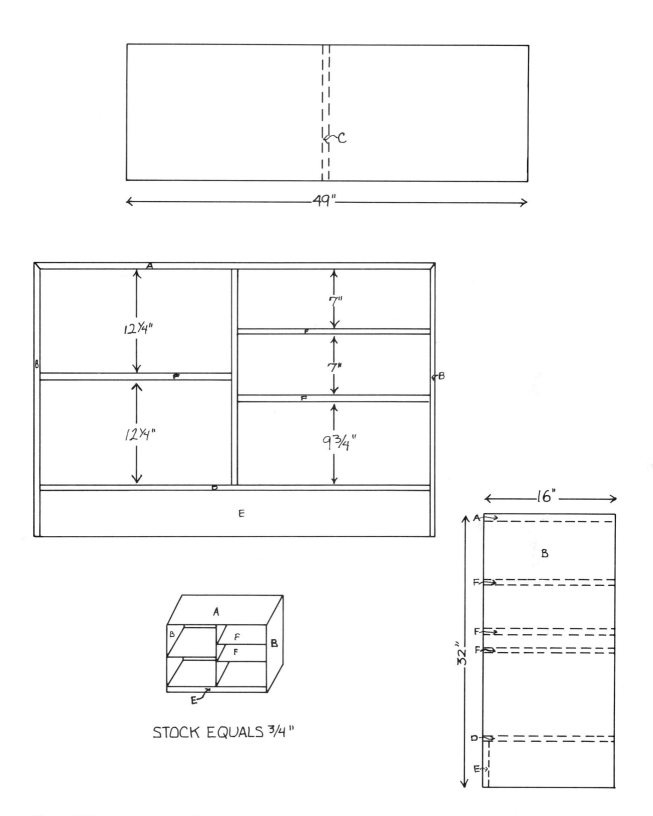

Illus. 197. Drawing detailing the parts and dimensions for the chest.

Cutting List

Piece	Quantity	Size	Description
A	1	16 × 49"	top
B	2	16 × 32"	sides
C	1	16 × 27¼"	center divider
D	1	16 × 47¼"	bottom
E	2	3¼ × 47¼"	bottom supports
F	3	16 × 23⅜"	shelves

After you have cut out and mitred pieces A and B (see the next chapter), lay them out for joining. I used seven biscuits on each piece to join pieces E to D, seven more on each end of this joined unit to attach it to the sides, five biscuits at either end of C to attach it to A and D, five biscuits on each end of the shelf to attach it into B and C, and five more on each end of A to mitre it to the corresponding pieces for B, for a total of 78 biscuits.

Before I had a joiner, I would never have considered mitring the top of the project in place. With the joiner, this job was easy.

This carcass unit only took approximately 90 minutes to cut out, join, and dry-assemble from the solid sheet. Taking it back apart and gluing it took another 25 minutes.

XVIII
Mitre Joints

You can also use the joiner to make boxes and other carcass projects with mitred rather than butted corners. Fine European-style furniture is mitred and splined together, and now these kinds of joints can be made in home workshops with biscuit joinery.

Generally, we think of mitred joints as those cut at 45 degrees, so we will treat mitred joints of other degrees in the

Illus. 198 (below). Side view of the carcass mitre joint. Illus. 199 (right). A long mitre joint in a carcass unit.

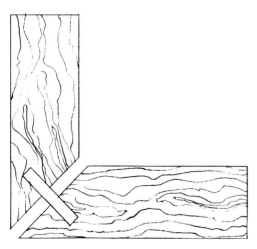

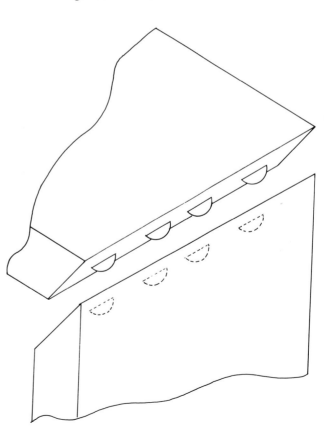

next chapter. The hardest thing about the mitre joint is preparing the stock for cutting the joint. Since the standard mitre gauge on a table saw isn't much good for crosscutting wide stock, clamp a straightedge accurately to the carcass piece and use the edge of the saw as a fence. This works well, but it provides ample possibilities for imprecision, and it is important that the pieces correspond to one another accurately if the joints are to work.

Following are the steps for preparing the stock for cutting:

1. Measure and mark (square) the length you wish to cut. If your setup is less than square and true, the results will be imperfect, but probably still better than when using the crosscutting fence alone.
2. Check the measurement from the edge of the saw table to the blade (on most saws, this is exactly 18 inches).
3. Mark 18¾ inches down from your cutting line (if you're cutting ¾-inch material; otherwise, add whatever is the thickness of the stock to the 18 inches) and clamp a straight board across the marks; check and double-check for square. I prefer a 2 × 2-inch board that is perfectly straight.
4. Remove the rip fence from the saw and make the cut using the edge of the saw table as a guide. This is much more accurate than using most mitre gauges for this kind of cut. You can square imperfect cuts by loosening one of the clamps and moving the batten just ever so slightly.

After you have made the cut (Illus. 200–202), and the pieces are ready for joining, mark the stock on the mitred faces, adjust the depth of cut, and cut the slots two inches from the edges and about four inches on center between them.

The pieces are now ready to be joined. First cut and fit any internal pieces for the biscuits, and then glue the whole unit (remember, only in the slots) and assemble it, preferably with a band clamp.

The Porter-Cable 555 is the only plate joiner that will accurately mitre together pieces of unequal thickness so that the outside edges meet without the need of a specialized setup. (See Illus. 203.) The "extra" end grain is on the inside of the joint where it is less likely to be visible. This way of mitring is substantially different from the approach taken by the other joiners. Other joiners, with the exception of the Lamello Top and, to a lesser extent, the Bosch model, have fixed angles at 90 and 45 degrees. (The Lamello and Bosch models are more or less continuously variable between 45 and 90 degrees.) All these joiners index their slotting cuts against the inside of the

Illus. 200. Cutting mitre joints with a Lamello Top.

Illus. 201 (above). Cutting a 45-degree joint with a fixed-angle joiner. Illus. 202 (right). It is better if you cut the pieces while they are firmly fixed in a vise.

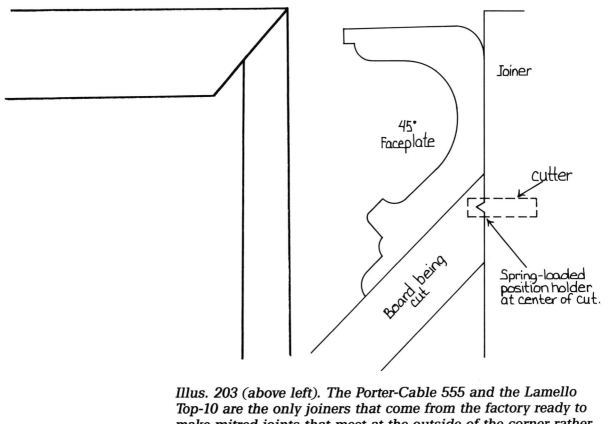

Illus. 203 (above left). The Porter-Cable 555 and the Lamello Top-10 are the only joiners that come from the factory ready to make mitred joints that meet at the outside of the corner rather than the inside. Illus. 204 (above right). This drawing shows the first cut by the Porter-Cable 555 or Lamello Top in what well may be a double-biscuited carcass mitre joint.

Illus. 205. Shown here is the standard method of cutting mitre joints.

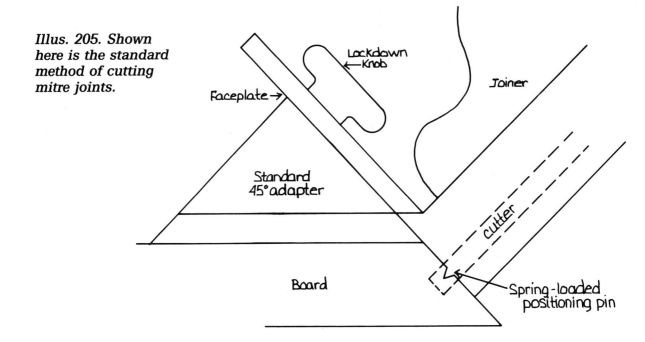

work. It is, therefore, fairly easy to cut either through the work or, worse, so close that exterior finishing makes the flaw visible. Also, unless the operator is meticulous, mitres cut with an inside-orientation mitre faceplate are likely to end up out of square somewhere in the joint; this is dangerous to the assembly, especially if you glue without first making a trial assembly.

I decided to build a simple compact disc storage unit, to hold my music collection until I could build a permanent unit. Below I describe the procedures for building this unit, which contains mitred corners. Bear in mind as you read this description that the entire project took less than two hours to build and finish.

The first step to building the unit is, of course, assembling the materials, which are as follow:

Piece	Quantity	Name	Size
A*	2	Top, Bottom	23 × 5¾"
B*	2	Sides	12½ × 5¾"
C	1	Shelf	21½ × 5"
D	3	Dividers	5 × 5"

*There should be a ¼" rabbet along one edge for back. The A and B pieces have mitred ends.

Illus. 206. The completed compact disc storage unit.

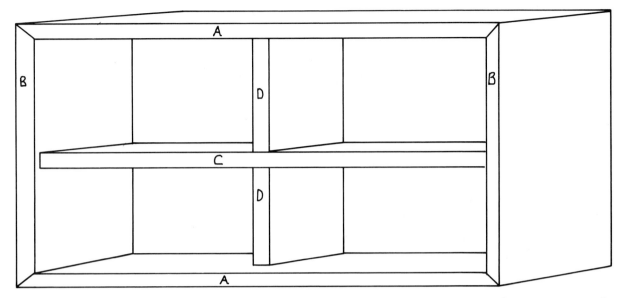

Illus. 207. Drawing of the compact disc storage unit.

After you have cut the pieces, test-fit the entire assembly. Sand all the inside pieces and the edges clean. Mark out all the joints and cut them. Cut the flat joints using the "paired" method already discussed. Cut the mitres all at once, using the joiner's mitre-cutting adjustment. Dry-fit the pieces and make any adjustments. Use the Lamello gluer; you will save an enormous amount of glue and cleanup time.

Use your imagination to adapt the unit to suit your own requirements.

XIX
Miscellaneous Joints

Cutting Offset Joints

Offset joints are all cut in the same manner as regular butted joints. Lay them out together, and then cut them with a shim in the joiner at one of the cuts. For example, the apron and the legs on a table sometimes aren't flush. You can set the apron back from the legs by simply setting a piece of ⅛-inch masonite or other appropriate shim between the apron and the height adjuster on the joiner's faceplate. (See Illus. 208–210.)

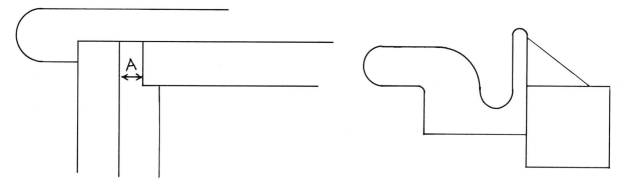

Illus. 208 (above left). A table-leg apron joint is a perfect example of an offset joint cut with a shim. Illus. 209 (above right). Cut the leg in the regular fashion.

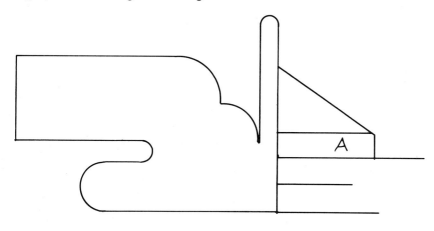

Illus. 210. With shim A equal to the offset shown in Illus. 208, cut the apron; on assembly, the fit will be offset perfectly.

141

Laying Out Narrow Work

You can lay out narrow pieces sometimes without even marking the work; the guide marks on the machine will show the outer limits of the piece and permit you to center the piece quickly using only your eyes. (See Illus. 211.)

Illus. 211. Laying out the narrow work with the guide marks on the bottom of the joiner.

Cutting Joints at Nonstandard Angles

Joints other than 45- or 90-degree joints can be cut almost as easily as 45- or 90-degree joints, especially if you use the Lamello Top, for it is continuously adjustable through 45 degrees. Cutting these joints is simply a matter of making an appropriate shim and fastening it to the fence of your joiner. Hot glue and duct tape work quite well for this fastening (Illus. 212), but you may want to screw these shims in place if you have many such joints to cut.

Illus. 212. Shown is all the equipment needed to mount the various angle blocks. Clockwise from the top are the following: duct tape, the shop-made angle blocks, the fixed-angle faceplate, and a glue gun.

The shims should be about the same size as the face of your fixed-angle faceplate, generally about 2 × 5 inches (the shims for the Lamello Junior model should be about six inches long), and can be attached to either the square or mitre face, whichever you find easier and more accurate. After you have removed the shim that has been glued and taped, you'll find that cleanup is surprisingly easy.

Illus. 213 and 214 shows the making of a nonstandard angle joint: A 15-degree shim is taped to the square side of the fence to make a 30-degree joint. While you could also tape it to the mitre face, doing it this way creates a "gap-proof" cut. This taped-on gauge was adjusted for this application simply by setting the front of it on the bench and then screwing it into place. This is a simple joint, but it can be used in a lot of applications.

Making Face-Frame Joints

Kitchen cabinetmakers, among other sheet stock users, will find the joiner useful for making the joint that attaches the hardwood face frame to the carcass, which was probably made of plywood. To set up for a face frame joint, make jig blocks as wide as the offset shoulder, as shown in Illus. 215. This block is the basic setup for the entire operation.

Make the cuts in the frame piece with the joiner resting on its base against this jig-block, as in Illus. 216, and then cut the face piece with the fence resting on the edge, as in Illus. 217. Illus. 218 illustrates the perfect centering that this technique delivers.

Illus. 213. To cut at a nonstandard angle, mount an angled shim on the bottom of the joiner.

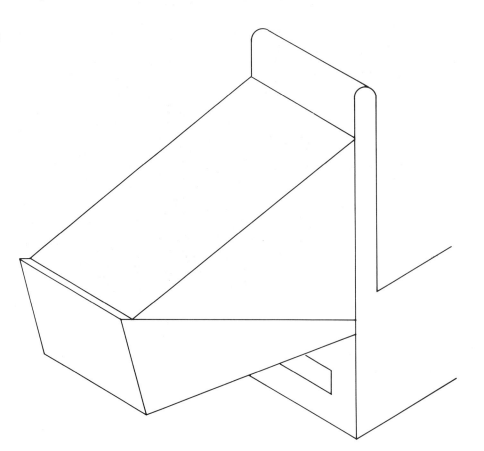

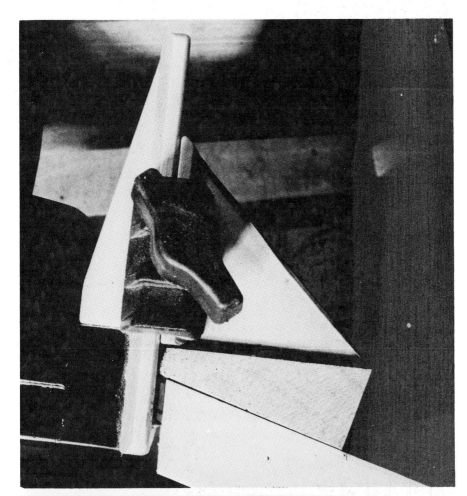

Illus. 214. Cutting at a nonstandard angle with a fixed-angle joiner and a shop-made shim. Note that the shim could be applied to the joiner either way, but the cuts tend to fit better when they are made this way.

Illus. 215 (below). Laying out a simple face-frame joint.

Illus. 216. Making the carcass end of the face-frame joint.

Illus. 217. Cutting the groove in the back side of the face-frame joint.

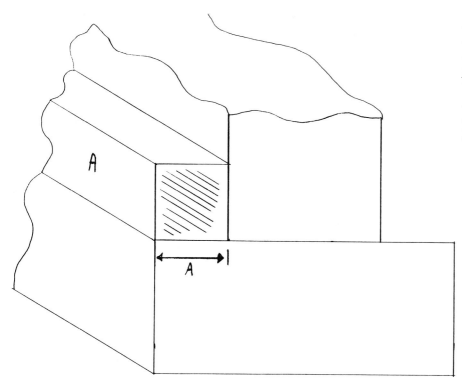

Illus. 218. Shim A is A inches square by at least the length of the joiner's faceplate width. The shim helps deliver the accurate centering that is needed.

If you make the face frames with the joiner, the narrowest stock you can use is 2¼ inches, for that is the narrowest piece that will hold a biscuit slot. There are other excellent ways to join narrower face-frame members. The new appliance, the FaceMaker, is probably the best. After a practice run or two, you'll find the method I've discussed to be superior to most of the other methods.

Projects

The following projects use joints discussed here and in previous chapters, and provide the perfect opportunity for you to test your joint-making skills.

MAKING A PICTURE FRAME

Making a frame can be simplified a great deal with the biscuit joiner if each piece is at least 1¾ inches wide so that the diagonal portions of the mitres will be 2½ inches wide; this allows them to be slotted for the biscuits without being scarred through the sides. Position the joiner carefully on this narrow work; cut slowly for precision. The slots will more or less fill these mitres. Cut the biscuit slots before cutting the rabbets which hold the picture and the glass.

Be sure to complete your sanding before assembling the unit; all you will be gluing will be the biscuits, since the glue on the end-grain-to-end-grain joint would be wasted anyway. Frame joints are at their strongest if they are double-biscuited, as shown in Illus. 219. The finished mitred frame will be much stronger and more attractive than one that has been just glued

Illus. 219. This frame joint has two biscuits, a number 10 biscuit nearer the front of the frame, and a number 0 biscuit nearer the back.

Illus. 220. After it has been biscuited together, the frame should be clamped with a band clamp like the one shown here; for a larger mitred joint, a more substantial clamping system might be desirable.

or glued and nailed. If the frame is of material large enough to accept even a size 0 biscuit, just a single biscuit in each corner is better than cross-nailing in terms of the strength of the joint and the ease of application. Of course, clamping the frame with a band clamp like the one shown in Illus. 220—even if only for ten or fifteen minutes—is a good idea.

MAKING A DRAWER RING

First, mark out the joint either freehand or with a rule. Next, cut slots with your joiner; if the material is thick enough to accept them, two or more biscuits per corner are advisable because they are so much stronger. Third, glue the biscuits and joint in place; it is easier to remove any excess glue before it has dried rather than after it has dried. There will be some cutthrough showing unless your side materials are at least 2¼ inches wide, but this really doesn't matter. Remove the clamp marks

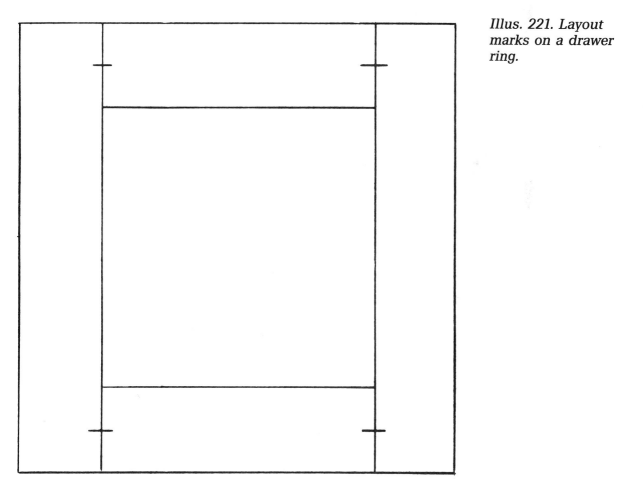

Illus. 221. Layout marks on a drawer ring.

from the "good" face if you forgot to use the protective clamp block.

The piece is now ready to be biscuited into place. Use the procedures for the internal carcass member on the drawer ring (pages 128 and 129).

XX
Shop-Made Accessories

Filed Fixed-Angle Faceplate

Since I am advocating cutting mitres and nonstandard joints with jigs set up to cut from the outside of the joint rather than the inside, thus preventing gaps in the joints, the easiest shop-made jig to use is the Porter-Cable fixed-angle frontplate, filed judiciously to fit standard joiners. Illus. 222 shows this plate fitted to a Lamello Top joiner. I prefer this much more for the Lamello Top than the Lamello fixed-angle frontplate.

Illus. 222. After a little judicious filing, the Porter-Cable fixed-angle faceplate can be fitted to the Lamello Top and, no doubt, other joiners. This inexpensive part might be a desirable attachment for your joiners.

A flap-front faceplate does hold an advantage over the fixed-angle frontplate: It can be set square without a square. The only ways to set a fixed-angle frontplate are either with a square or by positioning the tool on a surface with a "square" piece of the right size to measure against; in each case, the built-in square marker is inadequate. The prototype for the Porter-Cable fixed-angle faceplate has a built-in square marker. I have a joiner like this that is about $\frac{1}{32}$ inch out of square, probably because of a misplaced screw hole.

Perhaps a better approach than adapting Porter-Cable's mitring fixture to your flap-front machine is cutting out a $2 \times 2 \times 5$-inch, 45-degree shim and attaching it to the flap front with a couple of sheet-metal screws (through existing holes). With this jig, you can cut 0–45-degree mitres from the outside of the board rather than the inside. This shim is one of my favorite accessories for the Lamello Top. (See Illus. 223 and 224.)

Illus. 223. The Lamello Top with the shop-made angle-setting jig in place. Note that this jig may sometimes be preferable to the flap front in that it cuts the angled joints from the outside of the joint rather than from the inside.

Illus. 224. Here is the Lamello Top set up to cut mitres from the outside of the board at any angle. Now if only there were a way to move the hinge on the joiner's front up and down. . . .

Gauge Blocks

A set of gauge blocks for wood of the most common thicknesses will save you a great deal of time. Loosen the adjustable faceplate, and set it on the gauge with the base of the joiner flat on the bench. These gauges are better to use than the joiner's "square" slides because the slots on the mating pieces must be exactly parallel. They are especially useful with the fixed-angle faceplate for the Lamello Top or for any of the faceplates that are marked only in metric.

This set of blocks will also make it easier to set up the tool for stacking biscuits when you need extra mechanical strength in the joint, particularly when joining thicker stock.

The blocks should be about 2 × 7 inches so that they can be big enough to fully support the fixed-angle faceplate and still leave its label exposed; the label shows the measurement of the block. My set of blocks runs by sixteenths of an inch from ⅛ to ⅞ inch thick, with a block of Baltic birch plywood thrown in for good measure (my lumber dealer supplies this stock as ½ inch thick, but it isn't). The sizes can be combined for thicker pieces or you can choose various sizes when you want to stagger biscuits in a joint that will be heavily stressed. (See Illus. 225 and 226.)

Illus. 225. This partial collection of depth-adjusting blocks is shown in front of a joiner with a fixed-angle faceplate.

Illus. 226. A depth-adjusting block in place on a joiner with a fixed-angle faceplate.

Fixed-Angle Fence and Joining Table

The May/June, 1986 issue of *Fine Woodworking* features an article by Graham Blackburn called "Plate Joinery: It's Strong Enough for Chairs." From the perspective of one who has never made a chair, I found the most interesting feature of the article to be his description of a jig he made for his Virutex O-81 joiner. The standard joiner has a fixed-angle fence to control the location of the slot. This fixed-angle fence can be moved up and down, thereby allowing the blade to enter the work at varying points within the stock thickness, an absolute necessity when you want parallel rows of biscuits in a joint. The stand Blackburn fabricated for this unit was really just an extension of the tool's fixed-angle fence. The table was made of particle board covered with plastic laminate and was screwed right to the fence. The entire setup could be clamped to the bench top, so the work could be brought to the tool rather than the tool to the work. This simple adaptation made it possible to safely biscuit-join pieces that would otherwise be too small to hold securely.

With this setup, all sorts of shapes and sizes can be accommodated by clamping stops and blocks to the table. When the work is fed into the cutter against the machine's spring-loaded mechanism, the entire 12 by 16-inch table moves. To make angled joints, simply add wedges to raise or lower the workpiece's angle of approach to the blade before you clamp the workpiece to your table.

Blackburn's next step was to ensure a strong joint. He achieved this by using two biscuits per joint, positioning them side-by-side like twin tenons and thereby doubling the effective side-grain gluing surface. Since the plates would fit perfectly in their machined slots, the chances of a weak joint due to a poor fit were virtually eliminated. Almost all the joints he made were offset, unlike the flush-surface joints of typical face-frame work. Adjusting the position of the table and the fixed-angle fence to which it's attached took time, but, once done, the speed, accuracy, and ease with which the joints were cut repaid Blackburn handsomely.

When I inquired of Graham Blackburn whether I could have some photographs of this jig to share with you here, he reported that he no longer had the jig in question, having recently adopted the Porter-Cable 555 as his joiner of choice. He did report, though, that he missed the jig and would soon be buying another standard plunge-type joiner for exclusive use with such a jig. Thus, Illus. 227 and 228 show not his imple-

Illus. 227. The Blackburn-style table is shown here disassembled.

Illus. 228. The Lamello Top is mounted to the Blackburn-style table to biscuit small pieces.

mentation, but rather my very quick adaptation of his method. I made a 12 × 16-inch table as he did, routed a notch to fit the joiner's fixed-angle faceplate into, and determined where the screws would go to hold it in place; at that point I decided that a layer of sandpaper—about 120 grit—would be superior, for my applications anyway, to the plastic laminate that Blackburn used, at least in terms of "gripping" small pieces. (Illus. 227 and 228 show almost the entire setup. They do not show the screws through the faceplate.)

As I tried to reproduce his joining table, it occurred to me that a nearly ideal table for such a jig would be the INCA mortising table, as shown in Illus. 229, optional with the old-style INCA table saws. Preferring a Delta UniSaw to the smaller INCA saw, I bought the mortising table for independent mounting. Now that the joiner has just about made small-

Illus. 229. Before this INCA mortising table was used with the Lamello joiner, it was used with the mandrel shown and an assortment of router bits.

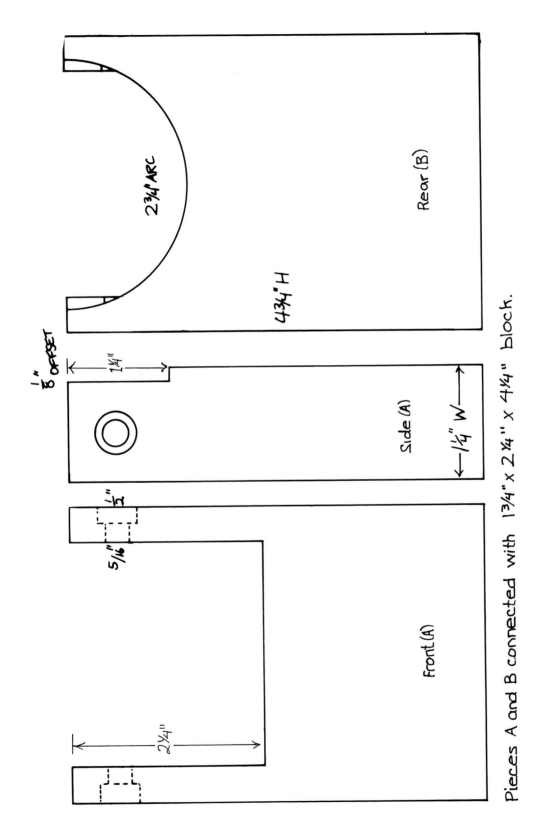

Pieces A and B connected with 1¾" × 2¼" × 4¼" block.

Illus. 230. The mounting platform for the Lamello Top.

scale mortise-and-tenon joining obsolete, I have adapted this table for slotting small pieces.

Nearly everything on the joiner is round, so one might at first be somewhat confused as to how to mount the table. However, it soon becomes obvious: Remove the handle and use it as a pattern for the front yoke of your table-mounting clamp. I like the handle off, as shown in Illus. 231, so much that I almost hesitate to put it back on; after all, it doesn't seem to serve much good.

Illus. 231. The Lamello Top looks better—and works at least as well—without its handle.

Full dimensions for the pieces to fit a Lamello Top joiner are given in Illus. 230. First drill the ¼-inch-deep recesses for the screw heads; then drill a ⁵⁄₁₆-inch hole all the way through the 3¼-inch width of the block. Make both cuts a half an inch from the top of the block.

After the drilling is accurately done, cut a channel 2⁹⁄₁₆ inch wide by 2¼ inch deep for the joiner to fit into; it's better and, in the long run, quicker, to file this channel out to width rather than leave the fit so sloppy that you have to start over. Next, drill the countersink and the hole for the ⁵⁄₁₆ × 6-inch bolt that will attach the piece to the table. After inserting the bolt, screw this yoke to the joiner with the handle fasteners and mount it to the table. This block is shown in Illus. 232.

For the second piece in this unit, saw nearly half of a 2½-inch-diameter circle from one end, and then cut it to length

Illus. 232. This is the main support block for the INCA table when it is used for biscuit joining rather than standard mortising.

at the other end so that the bottom of the joiner is exactly parallel to the INCA table. After cutting its countersink and bolt hole, mount it to the table. The unit is now ready to be used. (See Illus. 233–235.) I found it necessary to use a piece of card stock to shim one of the supports to ensure perfect square.

This unit is made in such a way that the INCA table can be used with either the joiner or the mortising head for which it was originally designed; the changeover time is under five minutes, assuming, of course, that the needed parts and wrenches are immediately at hand.

It may be instructive to compare this jig to the Lamello

stationary attachment used to mount the joiner to a commercial drill press's column. Of course, shopmade jigs aren't as sophisticated, but they aren't as expensive either. I mounted the joiner to the INCA table in under an hour with pieces of scrap wood; the net cost, excluding my labor, was zero.

Table-Model Biscuit Joiner

The shop-made table-model biscuit joiner isn't really an accessory, but rather a substitute for a joiner that was built in the days before a modest portable joiner could be bought inexpensively.

This table-model biscuit joiner is constructed by con-

Illus. 233. Underside view of the joiner/ INCA table setup.

Illus. 234. The most important thing about the setup is that the joiner's blade must be absolutely parallel to the INCA table.

Illus. 235. The joiner/ INCA table setup at work cutting slots in a piece of quite narrow picture frame material.

necting a motor to one end and a saw blade (one that makes a 4-mm kerf and has a 105-mm [just over 4 inches] diameter) to the other end of a pillow block mounted in a cabinet. The table is mounted on a pair of under-drawer drawer slides; a pair of guides hold it in place. Saw kerfs serve to mark the proper depth for size 20 biscuits; thin fillers could handily set up this table for size 10 and 0 biscuits. Sandpaper glued to the table holds the material steadily in position for slotting.

The cutterhead (Illus. 236) is covered with a piece of GE MARGARD plastic sheeting; this quarter-inch-thick clear plastic material is so hard it cannot be broken, even when struck with a pickax. It is exactly the sort of material you want between you and a failed cutter.

There are some problems with this joining table. First of all, there is no vertical adjustment. The cuts must be made at a predetermined height on the table. Second, there is no spring return to help prevent the person from losing his balance over the system while returning the table to its "start" position (Illus. 239 shows the operating pose this table requires). Finally, not all of us have access, as the fabricator of this table apparently does, to a neverending supply of scrap parts with which to build these tables. The man who built the table maintains that the box of biscuits he bought at discount represents fully half his cost for the device, including the safety features. By comparison, even the least expensive commercial joiners seems outrageously expensive.

Illus. 236. The cutterhead on the shop-made stationary biscuit-joining table.

Illus. 237. Another view of the shop-made biscuit-joining table.

Illus. 238. The outboard side of the shop-made biscuit-joining table. Note here the underdrawer glide and the cutouts that help to control the depth of cut.

Illus. 239. Operating the shop-made stationary joiner.

XXI
Projects

They say that a plumber's faucet always drips because the plumber has no time to make repairs at home. This saying can also apply to me and my woodworking shop; over the years I kept accumulating woodworking books and magazines that were too interesting to throw away. My shop became so cluttered I decided to build a fixed-shelf bookcase to store this material.

The main difference between building this bookcase with a joiner and building it by more conventional methods are those of time and convenience. Were I not building it with biscuit joints and mitred corners I would have to do the following: dovetail the upper and lower corners, add the base to the unit, and let the top shelf and center upright into router-cut dadoes in the underside of the top shelf. Cutting the ¼-inch × ¼-inch rabbet for the dust panel on the back would be more difficult, and I would have to rest the shelves on shelf standards.

Using the conventional method, I probably would have started without a formal plan, knowing that I needed a bookcase approximately four feet high by two and a half feet wide; the construction process would turn out to be more time-consuming, very frustrating, and not very efficient, though the results might have been very satisfying.

The joiner has brought efficiency to my shop. Here is the new, quicker order of construction with this tool. First, begin with a drawing; it doesn't have to be elaborate or to scale, but

Illus. 240. Drawing of bookcase.

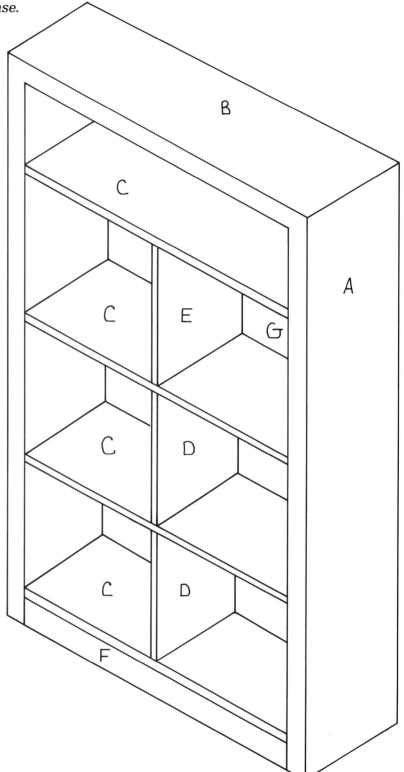

Cutting List for Bookcase

Pc	Qty	Name	Width	Length	Thickness
A	2	Sides	9½"	48"	¾"
B	1	Top	9½"	30"	¾"
C	1	Shelf	9½"	28½"	¾"
C	3	Shelves	9¼"	28½"	¾"
D	2	Dividers	9¼"	11½"	¾"
D	1	Divider	9¼"	12"	¾"
F	2	Kick plate	2½"	28½"	¾"
G	1	Back panel	29"	45½"	¼"
	46	Biscuits			
		Adhesive			
		Finishing material			

the dimensions must be right. Check and double-check your arithmetic as you make your cutting list. An accurate cutting list is absolutely essential to productive construction with joining biscuits.

Accurately cut all the pieces on the cutting list in your shop. (See Illus. 241.) If you are reproducing my bookcase, note that the bottom on one of the C members (Illus. 240) is ¼ inch

Illus. 241. You can cut off all the pieces exactly to length by attaching a stop to the crosscut fence of your table saw.

wider than the others. Cut the mitres for the sides first. Then cut the pieces to length with the mitres meeting; this will ensure sides of equal length. Set the crosscut gauge on your saw so that the piece fits exactly to the inside edges of the mitre joint before cutting the interior pieces to length.

Dry-assemble the pieces, adjusting them to get as near a perfect fit as possible. Make sure that the mitres you cut are square and true. Cut the rabbets for the back panel as you are cutting out the pieces; part of the beauty of joiner/biscuit construction is that you don't have to cut the rabbet by working all around the assembled carcass with a router.

Next, lay out the pieces for the biscuit slots that you will have to cut, always working in a logical order. The bookcase, for example, is best assembled on its side; after you have joined all pieces A and C in place, you can't join pieces D and E between them. The progression should work stepwise from bottom to top.

Because books are heavy, I used three biscuits per side on each shelf; two biscuits per end are sufficient to stabilize the spacers (C and D). Cut the slots with your joiner. The two kinds of cuts required here are for the carcass and mitre joints discussed in Chapters XVII and XVIII. Don't forget the single biscuit that joins each end of pieces F to pieces A; three or four more biscuits are enough to join F to the C you've selected as the bottom shelf. The layout takes just a few moments, and the cutting goes just as quickly.

After you have cut out the pieces and laid out and cut the biscuit slots, you may be tempted to assemble the piece as quickly as possible; don't. Before you glue the pieces together, do all but the final hand sanding; this will expedite things, for a 120-grit sanding belt can save time and produce good results much more quickly than 150-grit abrasive in an orbital sander.

After the sanding is completed, assemble the bookcase. First glue and biscuit pieces F to C; spread the glue into the slots, and run a line of glue between the slots on the surfaces to be joined. Use enough glue so that the surfaces join firmly, but not so much that you have a lot of glue to clean up. Some woodworkers prefer to remove the glue with a wet rag; I've always let it harden to at least the consistency of cottage cheese and then take it off with a razor-sharp chisel.

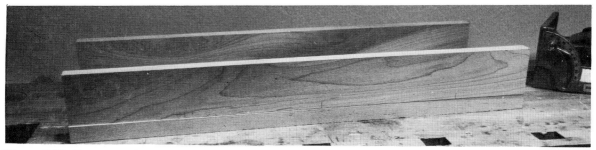

Illus. 242. Bottom shelf assembly.

Illus. 243. Bottom shelf ready for slotting.

Illus. 244. Here is the bottom shelf with the slots in place.

Illus. 245. The layout for the next shelf. This layout continues all the way to the top.

Illus. 246. The bookcase assembled.

Clamp these joints lightly and briefly. I've found that Weldbond Professional Woodworker's Glue is ready to machine in 30 minutes, and when biscuiting I can usually remove the clamps in 15 minutes or less. Spring clamps with a 3¼-inch capacity are adequate for this job.

Even with the clamps in place, you can proceed to glue, biscuit, and join pieces C and F to A, which you're using as the bottom. Then glue and biscuit a D and another C, and add them to the assembly. Next, add another glued and biscuited D and C, and then an E and the last C.

Glue, biscuit, and assemble the mitre against the bottom. Then prepare all the slots that will meet the "top" A piece. If you are working alone, it will take a bit of work to get the pieces to drop in perfectly, but if you have dry-fitted accurately, you will succeed. Using a carcass clamp, clamp the unit in place for a short while; if you are working and building a lot of carcass pieces, the Lamello spanner set will prove extremely valuable.

After allowing the glue to set, remove the clamps, scrape any excess glue away, tack the back panel into place, and complete your hand sanding in preparation for applying the chemical finish of your choice. I used two applications of Watco Oil and a coat of Goddard's Cabinetmaker's Wax on mine.

I built this bookcase so quickly that by Monday evening I had all my woodworking books and back issues of *Popular Woodworking*, *Fine Woodworking*, and *American Woodworker* stored away neatly. If a couple of hours will take us from raw boards to finished product, most of us are more likely to take on small projects for the home.

Small Table

A small table is always useful, and this one can be made in just one afternoon with only about five board feet of material. Since the project uses so little material and can be assembled so quickly, you can well afford to spend a few extra minutes choosing the right piece of solid or veneered stock for the top, and you should be able to shape a much more attractive lip for it than my drawing shows.

There are two principal advantages to making the table with the biscuit joiner. One, the time-consuming mortise-and-tenon joinery where the legs meet the apron is replaced by biscuits, which are actually just as strong. Second, instead of having to choose one of the relatively unsatisfactory conventional methods to attach the top, you have the option of using Lamello Lamex fasteners; these fasteners are effective and portable.

As with all biscuit-joined projects, first cut your stock to

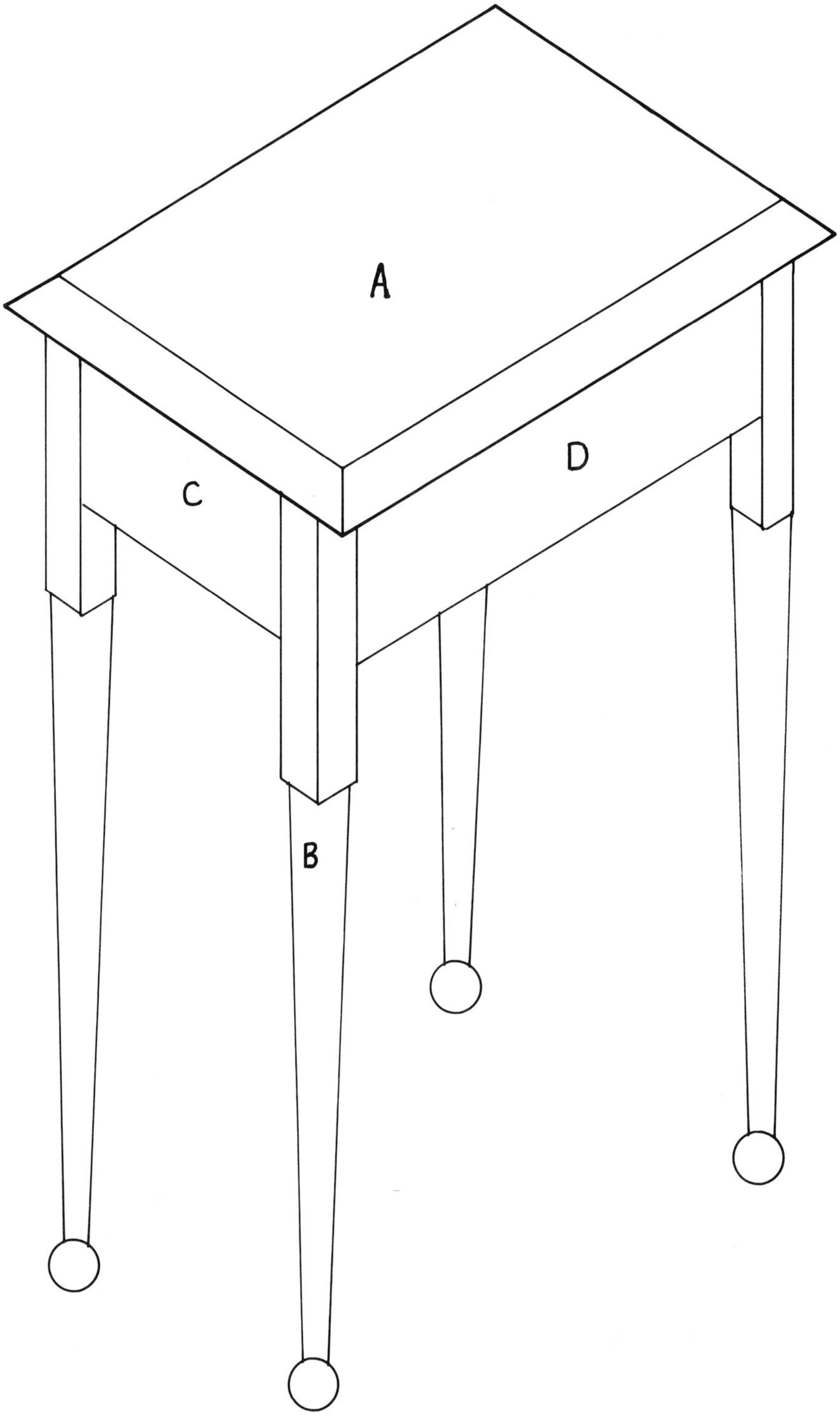

Illus. 247. Drawing of the small table.

Cutting List for Small Table

Pc	Qty	Name	Width	Length	Thickness
A	1	top	11"	16"	7/8"
B	4	leg	1½"	24"	1½"
C	2	apron, end	4¾"	6"	7/8"
D	2	apron, side	4¾"	11"	7/8"
	8	biscuits			
	4	Lamello Lamex biscuits			

exact size. The nine components of this table use so little material that you should be able to prepare the material very quickly.

The legs on this table are turned, a technique that isn't discussed in this book, so we will assume that any turning or carving will be part of stock preparation. It may be useful to note here that the first 7½ inches of each leg is 1½ inch square and tapers down from 1¼ to 7/8 inch, with a 1¼-inch ball on the bottom. While I might have cut my mortise-and-tenon joints before I completed other stages of stock preparation in the conventional method, joining is so foolproof that you can groove the stock after completing all preparations.

After the material is prepared, mark the legs and aprons for tandem, offset biscuits. Tandem biscuits double the amount of effective glue area, and offer an almost unbreakable completed joint. Cut first one pair of slots for each leg-apron joint, using a shim equal to the amount you want the apron to be inset from the leg's front as you cut each apron piece; then readjust your joiner's faceplate so that the second row of slots will be about ¼ inch in from the first row, and proceed to cut the next round of slots, again using the shim on the apron pieces.

Don't glue the legs and apron pieces together until after all but final hand-sanding has been completed. Then glue generously in the slots and very sparingly everywhere else. This leg set is almost too small to be clamped with a framing clamp like the Lamello spanner set, but it can be clamped quite successfully with small standard bar clamps. After the glue has set, remove the clamps and carefully scrape any excess glue away with a sharp chisel. Then, do the finish sanding.

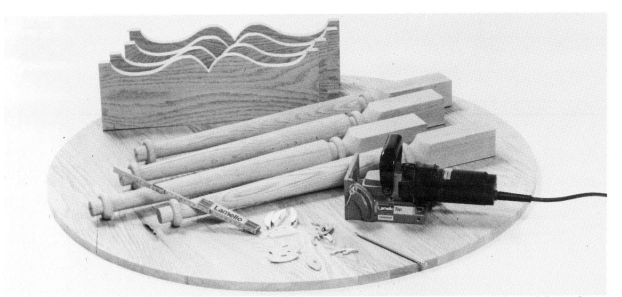

Illus. 248. Lay out the top, legs, and aprons of the table to be assembled.
(Photo courtesy of Colonial Saw)

Illus. 249. Select the inside joining surfaces on the legs and mark the center lines for the joining grooves.
(Photo courtesy of Colonial Saw)

Illus. 250. Repeat this process for the aprons using the same measurements.
(Photo courtesy of Colonial Saw)

Illus. 251. Using the center alignment mark on the Lamello machine, plunge-cut the grooves in the aprons.
(Photo courtesy of Colonial Saw)

Illus. 252. Then, while resting the joiner on the apron, plunge-cut the grooves in the legs, again using the center alignment mark.
(Photo courtesy of Colonial Saw)

Illus. 253. For a permanent apron/leg assembly, apply water-based glue to the grooves.
(Photo courtesy of Colonial Saw)

*Illus. 254. Assemble
the Lamello joining
plates.*
*(Photo courtesy of
Colonial Saw)*

*Illus. 255. Then as-
semble the aprons
and legs on a flat
surface. The plates
can be laterally ad-
justed for perfect fit.
Next, clamp using
the Lamello span-
ners and allow the
completed base to
sit for 20 minutes.*
*(Photo courtesy of Colonial
Saw)*

To attach the top with the Lamello Lamex, set the top on
the leg assembly and mark out the sides (long apron pieces)
for a pair of biscuit joints. Cut the slots in the top and in the
aprons. Glue the Lamex joining biscuits in the aprons. Following the instructions in Chapter XIV, make the multiple plunge
cuts to mill the wide slots for the Lamex fittings 90 degrees
from the existing slots. Use the Lamello hot-melt glue gun if

you have the Lamello Lamex kit, or epoxy them in place if you are milling these slots freehand.

After the Lamex connectors have been glued in place, attach the top simply by setting it in place and turning the four screws clockwise. Of course, you can use standard top-fastening techniques, but after you have tried the Lamex system, you will agree that it is a superior joining method.

After the top has been attached, give the project one last hand-sanding with very fine paper (220 grit if you have sandpaper that is that fine) and apply the finish of your choice.

Excluding the scrub-waxing that was done a couple of days after completion, this project went from design to gratifying completion in a single afternoon.

Illus. 256. Position the base in the underside of the top. Then mark the inside and outside perimeters with a pencil line. Also mark lines for the joining plates on both the apron and top on the inside surface of the two adjacent sides and on the outside surface of the two opposite adjacent sides. Next, slide the table base back and to the side so that the outside perimeter line is aligned with the inside perimeter of the two marked adjacent sides. Clamp or block the table in place and, as this photograph shows, plunge-cut the grooves in the top, using the inside of the apron as your guide and aligning the center mark on the top with the centering mark on the joiner. Repeat the process using the outside of the opposite apron to cut the remaining grooves.
(Photo courtesy of Colonial Saw)

Illus. 257. Using the right-angle plate on the Lamello machine, plunge-cut the grooves in the aprons at the center marks. Then machine the grooves in the apron using the inside as a guide on two sides. (Photo courtesy of Colonial Saw)

Illus. 258. Insert the Lamello Lamex joining plates with their holes exposed along the edge of the apron and aligned with the center marks. Allow them to set for 30 minutes. (Photo courtesy of Colonial Saw)

Illus. 259. To install Lamello Lamex K-D fittings in the top, place the Lamex cutting guide in the groove, place the Lamello joiner in the cutting guide, and make multiple plunge cuts to mill the wide slots for the fittings. (Photo courtesy of Colonial Saw)

Illus. 260. Press the fitting in place in the slot and glue with a hot-melt gun. In five minutes the table, as shown here, will be ready for final assembly and finishing.
(Photo courtesy of Colonial Saw)

Illus. 261. Mate the base to the top and tighten the Lamex fittings. Your table is complete. Assembly with biscuits was easier and faster than by any other method, and the result is a stronger table.
(Photo courtesy of Colonial Saw)

Appendix A
Product Update

The Bosch GUF 4-22A

The German-made Bosch GUF 4-22A joining and trimming saw will be made at Bosch's New Bern, North Carolina, plant after it has passed the UL test and been introduced to the United States market. This 620-watt, 10,000-RPM unit weighs just over five pounds. Its cord is 6 feet long. Like most of the other portable joiners, this one comes in an attractive and convenient case. A dust bag is standard issue, and an Air Sweep™ dust extractor is available at a fraction of the cost of the dust extractors for other units. (See Illus. 262 and 263.) Since the unit is a trimming saw as well as a joiner, there are several types of blade available, ranging from 12 to 22 teeth in thicknesses of 2.2 to 4 mm.

This unit is more similar to the Elu joiner than to any of the others. Like the Elu, its blade is adjustable along the edge of the tool. Both joiners feature superior switch action, but, unlike the switch on the Elu joiner, the Bosch switch can be locked on for some kinds of cutting. Like the Elu joiner, this unit's "angle-in," rather than "plunge-cutting," action allows for easy edge-trimming.

The models also have differences of interest to the prospective buyer: The Bosch joiner is likely to be less expensive than the Elu. Further, its fencing systems appear superior in a couple of important ways. For example, the mitre gauge can be set anywhere from 0 to 90 degrees. Also, the Bosch joiner features a spindle-lock rather than a "two-wrench" blade-changing system. While the fencing system isn't made of the highly machined (and, thus, very expensive) material of the Elu's fencing system, it is of high quality, and a removable strip of live

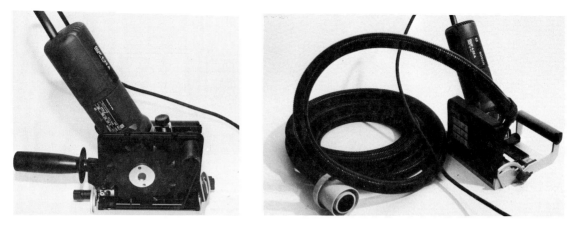

Illus. 262 (left). The Bosch joiner with its blade guard removed. Illus. 263 (right). The Air-Sweep™ system in place.

rubber helps keep the unit accurately positioned when biscuit-joining. Besides the removable mitre gauge, there is a sturdy rip fence, gauged in millimetres. Though of all the models I tested the Elu joiner did the best job of throwing the chips away from the left-handed operator, the Bosch joiner holds an advantage because it is fitted with a dust bag, and because there is a vacuum attachment for the machine for those who want a completely dust-free environment.

The Bosch model makes adjustments more easily than any of the other joiners. The depth-of-cut adjustment nut has a top-of-machine scale which is marked for biscuit size as well as for actual depth.

The Bosch GUF 4-22A will do edge-trimming to a depth of ⅞ inch with the biscuit-slotting blade or with a very fine 22-tooth (2.2-mm) blade. (See Illus. 264.) The combination of this

Illus. 264. The two cutting blades for the Bosch joiner supplement the biscuit-joining blade.

fine blade and the joiner's small size, handy rip fence, and generally well designed body make this machine much handier than a standard portable circular saw for edge-trimming.

This joiner also has some disadvantages. As has been mentioned, it and the Elu model cut at an angle rather than plunge as do the other units tested. Because of this, the manufacturer recommends that the operator always cut from the same side since the housing can't be marked for centering cuts, as can the housing on other units. This means you will either have to make some test cuts, fit the biscuits in continuous (rather than biscuit-size) slots, or lose the real convenience of the "standard" plate joiner. Being unalterably opposed to the idea of frequent test cuts, I think there should be a way to mark the housing so that "flip-flop" biscuit cuts could be made at least for size-20 biscuits; perhaps this marking could be done more accurately at the factory by the manufacturer.

If all you plan to do is plate-joining and your real goal is to save the maximum amount of time by doing so, this may not be the machine for you. However, since you probably haven't tested all the available plate joiners, the disadvantages mentioned probably won't concern you. Furthermore, this machine's solid German quality and its other advantages may well offset any disadvantages in your shop.

Product managers at Bosch believe that this joiner will be used in the flooring trade because it has enough blade extension to cut trim and because the fence can be used to trim a door parallel without the door having to be removed. There are sure to be people buying this joiner who will never use it for biscuit-slotting. If you plan on using your joiner for more than just biscuit-joining, this is probably the tool for you.

Woodhaven's Biscuits and Splines

Woodhaven supplies 1 × 1⁵⁄₁₆-inch stamped biscuits and 1 × 8-inch splines made of particleboard which has been compressed and then sanded to an exact ¼-inch thickness. The advantage of these biscuits is that they can be used to assemble face-frame units where the members are just 1½ inches wide. (See Illus. 265.) Another advantage is that router owners

Illus. 265. Wood-haven biscuits in place in a very narrow frame.

who like this system now have a setup that incorporates all the advantages of a biscuit joiner without having to buy one.

The router uses what appears to be a standard ball bearing ¼-inch slotting cutter. (See Illus. 266.) This bit seemed lost in the half-inch collet of the Makita router that's in my Woodhaven router table. To get enough elevation on the bit to do the slotting at the center of a ¾-inch board, I found I had less of the bit's shank in the collet than I would really have preferred. Much of this problem would have been eliminated by the use of a router that used quarter-inch bits, but since not all shops have more than one router, providing a slotting cutter that would work nicely in a half-inch collet might be a major improvement for this accessory.

The Woodhaven manual advises the operator to make the slots by pivoting the work into the bit. On one trial cut, I fed the work straight in rather than off a pivot, and I was rewarded for my carelessness with a smashed knuckle. However, this accident was a result of nervousness more than anything. Routers make so much noise that even though I was wearing hearing protectors I was momentarily frightened. However, if

Illus. 266. Cutting slots with a Wood-haven slotting cutter on a Woodhaven router table.

Illus. 267. Wood-haven slotting cutter (left) and the "standard" slotting cutter from my shop.

you like to use routers, or if you have a need to join pieces as narrow as 1½ inch, this may be the tool for you. (See Illus. 267.)

Sears Craftsman Bis-Kit System

Sears' Bis-Kit plate/edge joiner system will also permit you to sample biscuit joinery without having to buy a joiner. The Bis-Kit accessory fits on almost any router. (See Illus. 268.) Assembly yields a very sturdy product, though mostly plastic, that will permit you to make biscuited edge-joints with a router. (See Illus. 269.) The Bis-Kit takes Woodhaven's "Biscuits & Bits" concept to its logical conclusion. The three-tooth carbide-tipped wing cutter can enter the work only on or very near its edges. All adjustments are done with the router's depth-of-cut

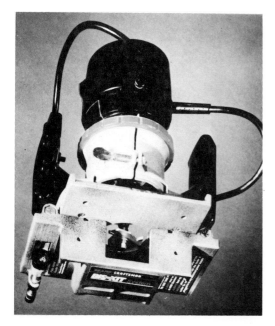

Illus. 268 (left). Router equipped with Sears Bis-Kit system. Illus. 269 (above). The Bis-Kit accessory in action. (Photos courtesy of Sears, Roebuck and Co.)

gauge. Sears' manual is a model of clarity, which sets it apart from all other joiner manuals.

This router accessory can be attached in about five minutes. That's fine if you already have a router and expect to do little biscuit-joining; otherwise, buying an inexpensive joiner probably makes more sense.

Use tests show that the unit does perform as specified. If you don't really plan to do lots of biscuit joinery and you like routers, this attachment will have you biscuit-joining for minimal expense.

The Porter-Cable 555

The Porter-Cable 555 has been improved a great deal by the redesign of the removable/adjustable 90- and 45-degree faceplate. The faceplate originally had to be set with a square each time it was adjusted, for there weren't enough large square edges to ensure perfect alignment. Now the faceplate wraps around the sides of the fixed faceplate and rides the edges of the fixed faceplate squarely. (See Illus. 270.) A good machine has been made better.

Wolfcraft Multi-Shaper

In Europe, the four-inch grinder is used nearly as often as the electric drill. European homeowners, do-it-yourselfers, and craftspeople have far more use for the small grinder than do Americans. In fact, most American woodworkers have never even used one. All this may change, for Wolfcraft of America, the United States arm of Robert Wolff GMBH, a major German tooling company, has introduced the Multi-Shaper, a device that will perform all the standard biscuit-cutting operations while powered by such a grinder.

The Multi-Shaper is made of fibre-reinforced acrylic, which gives it rigid strength and makes it inexpensive. It comes with a dust bag. Its instruction manual contains only illustrations, but they are sufficient to aid both assembly to your grinder and the operation of the completed "joiner." While details about the tool's United States availability and pricing weren't available at the time this book was published, it is

Illus. 270 (left). The new 45/90-degree front piece for the Porter-Cable 555 joiner.

Illus. 271 (above). Front view of the Wolfcraft Multi-Shaper.

worth noting that this has been Wolff's best-selling product in W. Germany since its introduction. (See Illus. 271.)

CabinetMaster CM100

Colonial Saw has recently introduced the CabinetMaster CM100, which uses milled slots and biscuits rather than dowels to achieve perfectly aligned joints and flush ends in Melamine-faced panels. (See Illus. 272.) The manufacturer claims that this product was designed for small- to medium-

Illus. 272. The electro-pneumatically controlled Cabinet-Master CM100.
(Photo courtesy of Colonial Saw)

sized shops, and will enable a single operator to mill a complete cabinet every two minutes and produce an average kitchen order in about an hour. All sequencing is electro-pneumatically controlled. The machine operates on 110 volt, single-phase service.

Jiggit Joiner Base

Jiggit has designed a base to fit Freud, Virutex, and Lamello joiners for use in stationary mode. (See Illus. 273.) This inex-

Illus. 273. The Jiggit Base in place on a Freud portable joiner. (Photo courtesy of Gaylord Livingston, Jiggit Mfg.)

pensive commercial base may make more sense than the shop-built base I described in Chapter XX on page 156 (Illus. 227 and 228). As with the shop-made unit, this base is attached by removing the handle so that those openings on the grinder body can be used to fasten the tool in its upside-down position.

Author's Note: Some expert joiners say biscuit joints should be used only in man-made sheet materials, but that is a matter of taste rather than practicality. So many handsome joints can be used with solid-wood construction: You might question whether to mitre and biscuit-join the corners of a solid walnut chest when through-dovetails would be so handsome. On the other hand, there is no better way to join the corners of a walnut-veneer chest. Modern equipment will cut the mitres accurately, and biscuit-joining is quick, accurate, and inexpensive.

Appendix B

Chart of Plate Joiner Features

Maker	Model	Address	Power Consumption	No load speed	Weight lb-oz	Warranty
Porter-Cable	555	Porter-Cable Corp. Youngs Crossing at Hwy 45 P.O. Box 2468 Jackson, TN 38304-2468	—	8,000 RPM	under 5	12 months
Freud	JS-100	Freud 218 Feld Ave. High Point, NC 27264	550 watts	10,000 RPM	6-3	12 months
Virutex	O-81	Rudolf Bass, Inc. 45 Halladay St. Jersey City, NJ 07304	500 watts	10,000 RPM	7	12 months
Elu	3380	Black & Decker, Inc. 10 North Park Drive P.O. Box 798 Hunt Valley, MD 21030	600 watts	7,500 RPM	7	12 months
Lamello	Junior and Standard	Colonial Saw 100 Pembroke St., P.O. Box A Kingston, MA 02364	500 watts	10,000 RPM	6-6	6 months
Lamello	Top	as above	600 watts	8,000 RPM	7	6 months
Lamello	Top-10	as above	700 watts	10,000 RPM	7	6 months

Maker	Model	Distributor				
Kaiser	Mini 3D	W. S. Jenks & Son 1933 Montana Ave. NE Washington, DC 20002	600 watts	10,000 RPM	under 6-8	12 months
Delta	32-100	246 Alpha Dr. Pittsburgh, PA 15238	1200 watts	10,000 RPM	39½	24 months
Bosch	GUF4-22A	New Bern, North Carolina	620 watts	10,000 RPM	5-4	12 months

Maker	Model	Manual	Sound Level No Load/Load	Case	Country of Origin	Cord Length feet/inches	Accessories
Porter-Cable	555	adequate	95/93 dB	metal	USA	7'	not yet announced
Freud	JS-100	poor	105/103 dB	injection-moulded	Spain	7'5"	not yet announced
Virutex	0-81	acceptable	101/104 dB	injection-moulded	Spain	7'3"	dust collector extra blade
Elu	3380	poor	97/103 dB	metal	W. Germany	10'7"	*
Lamello	Junior	poor	101/102 dB	cardboard	Switzerland	8'2"	dust extractor hinges centering awl
Lamello	Top	excellent	100/102 dB	wood	Switzerland	8'3"	See Chapter XIV
Lamello	Top-10	poor	93/93 dB	wood	Switzerland	8'	See Chapter XIV
Kaiser	Mini-3D	not available	98/96 dB	metal with wood liners	W. Germany/ Austria	almost 10'	metre vacuum horse
Bosch	GUF4-22A	not available	92/91 dB	metal	W. Germany	6'	Sweep™ dust extractor**

*Special features include grooving and scribing
**Special features include edge-trimming

METRIC EQUIVALENCY CHART

MM—MILLIMETRES   CM—CENTIMETRES

INCHES TO MILLIMETRES AND CENTIMETRES

INCHES	MM	CM	INCHES	CM	INCHES	CM
⅛	3	0.3	9	22.9	30	76.2
¼	6	0.6	10	25.4	31	78.7
⅜	10	1.0	11	27.9	32	81.3
½	13	1.3	12	30.5	33	83.8
⅝	16	1.6	13	33.0	34	86.4
¾	19	1.9	14	35.6	35	88.9
⅞	22	2.2	15	38.1	36	91.4
1	25	2.5	16	40.6	37	94.0
1¼	32	3.2	17	43.2	38	96.5
1½	38	3.8	18	45.7	39	99.1
1¾	44	4.4	19	48.3	40	101.6
2	51	5.1	20	50.8	41	104.1
2½	64	6.4	21	53.3	42	106.7
3	76	7.6	22	55.9	43	109.2
3½	89	8.9	23	58.4	44	111.8
4	102	10.2	24	61.0	45	114.3
4½	114	11.4	25	63.5	46	116.8
5	127	12.7	26	66.0	47	119.4
6	152	15.2	27	68.6	48	121.9
7	178	17.8	28	71.1	49	124.5
8	203	20.3	29	73.7	50	127.0

YARDS TO METRES

YARDS	METRES	YARDS	METRES	YARDS	METRES	YARDS	METRES	YARDS	METRES
⅛	0.11	2⅛	1.94	4⅛	3.77	6⅛	5.60	8⅛	7.43
¼	0.23	2¼	2.06	4¼	3.89	6¼	5.72	8¼	7.54
⅜	0.34	2⅜	2.17	4⅜	4.00	6⅜	5.83	8⅜	7.66
½	0.46	2½	2.29	4½	4.11	6½	5.94	8½	7.77
⅝	0.57	2⅝	2.40	4⅝	4.23	6⅝	6.06	8⅝	7.89
¾	0.69	2¾	2.51	4¾	4.34	6¾	6.17	8¾	8.00
⅞	0.80	2⅞	2.63	4⅞	4.46	6⅞	6.29	8⅞	8.12
1	0.91	3	2.74	5	4.57	7	6.40	9	8.23
1⅛	1.03	3⅛	2.86	5⅛	4.69	7⅛	6.52	9⅛	8.34
1¼	1.14	3¼	2.97	5¼	4.80	7¼	6.63	9¼	8.46
1⅜	1.26	3⅜	3.09	5⅜	4.91	7⅜	6.74	9⅜	8.57
1½	1.37	3½	3.20	5½	5.03	7½	6.86	9½	8.69
1⅝	1.49	3⅝	3.31	5⅝	5.14	7⅝	6.97	9⅝	8.80
1¾	1.60	3¾	3.43	5¾	5.26	7¾	7.09	9¾	8.92
1⅞	1.71	3⅞	3.54	5⅞	5.37	7⅞	7.20	9⅞	9.03
2	1.83	4	3.66	6	5.49	8	7.32	10	9.14

A. To convert Fahrenheit to Celsius:

Subtract 32 from the Fahrenheit degrees and multiply by 5/9.

Formula: (F. Degrees $-$ 32) $\times \dfrac{5}{9}$ = C. Degrees

$$\dfrac{5}{9}(F. - 32) = C.$$

B. To convert Celsius to Fahrenheit:

Multiply the Celsius degrees by 9/5 and add 32.

Formula: (C. Degrees $\times \dfrac{9}{5}$) $+$ 32 = F. Degrees

$$\dfrac{9}{5}C. + 32 = F.$$

or: 1.8C. $+$ 32 = F.

Index

ACKNOWLEDGMENTS

Special thanks to *Popular Woodworking* magazine, and editor, David Camp. Additionally, the book could not have been completed without the help of a number of folk from the plate joiner manufacturing and distribution business: Mr. Rick Schmidt and Mr. Dennis Huntsman, Porter-Cable Tools, Jackson, TN; Mr. Gary Compton and Mr. Fred Göbel, Robert Bosch, Inc., New Bern, NC; Mr. Barry Dunsmore, Freud USA, Inc., High Point, NC; Mr. Nick Basile, Rudolph Bass, Inc. (Virutex), Jersey City, NJ; Mr. David Myers and Mr. Jim Roberts, Black & Decker Corp. (Elu), Hunt Valley, MD; Mr. Sandor Nagyszalanczy, *Fine Woodworking* magazine, Newtown, CT, for the loan of his Kaiser joiner; and Mr. Bob Jardinico, Colonial Saw, Inc. (Lamello), Kingston, MA, who provided some of the photos that illustrate the text.

Special thanks to John Anderson and Bill Flemming, who helped with the technical drawings, and to Mike Cea, who turned my manuscript into a book. Like the representatives of the tool companies mentioned above, each person deserves more of a thank-you than appears here, for the book reflects the expertise which they have most generously shared with me. Of course, responsibility for any errors and omissions remains my own.